An Introduction
to Nervous Systems

Also by Ralph J. Greenspan

Fly Pushing: The Theory and Practice of Drosophila *Genetics*, Second Edition
Invertebrate Neurobiology (with Geoffrey North, to be published June 2007)

Other Titles of Interest

Imaging in Neuroscience and Development: A Laboratory Manual
From **a** *to* α: *Yeast as a Model for Cellular Differentiation*
A Genetic Switch, Third Edition: Phage Lambda Revisited

An Introduction
to Nervous Systems

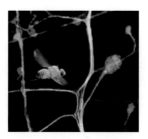

Ralph J. Greenspan

The Neurosciences Institute
San Diego, California

COLD SPRING HARBOR LABORATORY PRESS
Cold Spring Harbor, New York

An Introduction to Nervous Systems

© 2007 by Cold Spring Harbor Laboratory Press, Cold Spring Harbor, New York
Printed in the United States of America

Publisher	John Inglis
Editorial Director/Acquisition Editor	Alexander Gann
Book Development, Marketing and Sales Director	Jan Argentine
Developmental Editor	Michael Zierler
Project Coordinator	Maryliz Dickerson
Permissions Coordinator	Carol Brown
Production Manager	Denise Weiss
Production Editor	Kathleen Bubbeo
Copy Editor	Rena Steuer
Desktop Editor	Susan Schaefer
Cover Designer	Paula Goldstein

Cover and title page artwork: Main Image: Sensory (*red*) and central nervous system (*green*) neurons in the body wall of a leech. Reprinted from Jellies J. and Johansen J., *J. Neurobiol. 27:* 310, © John Wiley & Sons Inc. Fly: A female fruit fly, *Drosophila melanogaster,* glued to a fine steel pin with a droplet of UV-cured epoxy. Photo courtesy of Mark Frye. Reprinted from Huang G.T., *J. Exp. Biol. 206:* 785, © 2003 Company of Biologists.

Library of Congress Cataloging-in-Publication Data

Greenspan, Ralph J.
  An introduction to nervous systems / by Ralph J. Greenspan.
     p. cm.
  Includes bibliographical references and index.
  ISBN 978-0-87969-757-0 (hardcover : alk. paper) -- ISBN
978-0-87969-821-8 (pbk. : alk. paper)
  1. Nervous system. 2. Comparative neurobiology. I. Title.

  QP361.G67 2006
  612.8'043--dc22

2006038174

10 9 8 7 6 5 4 3 2 1

All Cold Spring Harbor Laboratory Press publications may be ordered directly from Cold Spring Harbor Laboratory Press, 500 Sunnyside Blvd., Woodbury, New York 11797-2924. Phone: 1-800-843-4388 in Continental U.S. and Canada. All other locations: (516) 422-4100. FAX: (516) 422-4097. E-mail: cshpress@cshl.edu. For a complete catalog of all Cold Spring Harbor Laboratory Press publications, visit our World Wide Web site http://www.cshlpress.com/.

For my lovely wife, Dani

Contents

Preface

Who are the invertebrates? Most of our early encounters with them are not very friendly. Jellyfish were something to avoid brushing against when swimming in the ocean; flies were good for swatting with a fly swatter on lazy summer days; leeches were to be feared when wading into muddy ponds (with the folk warning not to pull them off but rather to rub salt on them so that they would shrink); and protozoans were the reason to avoid drinking brackish water when hiking in the woods. In school, invertebrates showed up in high school biology class as those exotic-looking creatures whose names and taxonomy had to be memorized, and some of which (such as starfish or sea anemones) might become class pets.

Little did I suspect that someday I would be proselytizing the virtues of these organisms for understanding brains and behavior. That is one of the reasons for this book. Another is to introduce the interested reader to the richness of behavior in the invertebrate world and to the fascination of trying to understand those behaviors in terms of nervous systems. Yet another reason is to tantalize would-be biologists with the prospect of using the ever-expanding tool kit of genome sequences, molecular probes, genetic engineering, and imaging technologies, together with clever ideas for experiments, to explore the behaviors and neural mechanisms of organisms that would have been inaccessible to such studies in the past. And, finally, I wish to argue that understanding the brains and behavior of these creatures is valuable not only for its own sake, but also for understanding our own brains and behavior.

The strategy of the book is to start with simpler organisms and behaviors and end with the more complex. Each of Chapters 1–4 introduces a key principle of nervous system function (electrical signaling within a cell, chemical signaling within and between cells, nerve circuitry, and modulation of neuronal activity) in the context of a particular organism's behavior (swimming and avoidance in *Paramecium*, responses to light in barnacles and horseshoe crabs, coordination of swimming in jellyfish and leeches, and the modification of brain activity by experience in the sea snail). Chapters 5–7 then explore how these basic mechanisms are used in the brain of the fruit fly to organize more complex patterns of behavior (circadian rhythms, flight, and courtship), and Chapter 8 deals with the question of learning and cognition in the fruit fly, the honeybee, and the cockroach. In each chapter, ancestral origins or natural variations (to the limited extent we know either of them) are also discussed in an effort to place the mechanisms in their evolutionary context. Finally, Chapter 9 summarizes the key concepts that have come up in the course of the book and highlights some of the most tantalizing, unsolved problems in neurobiology.

In trying to introduce nervous systems and neurobiology, I have written this book for those who have the sort of background provided by an introductory biology course. Familiarity with the basic ideas of cells, membranes, proteins, DNA and RNA, enzymes, receptors, ions, and small molecules is assumed.

I would like to thank the following colleagues for lending their expertise and commenting on portions of the manuscript: Michael Dickinson, Jean-François Ferveur, Martin Giurfa, Dick Horn, Bill Kristan, Ching Kung, Bambos Kyriacou, Irwin Levitan, Bob Meech, Peter O'Day, Rick Ordway, Ann Stuart, and Bill Wright. Any errors that remain are solely my responsibility. I would also like to thank Esther Goldstein and Aubrie O'Rourke for their comments, criticisms, and suggestions. Any lack of clarity or readability that remains is solely my responsibility. In addition, I would like to thank Alex Gann of Cold Spring Harbor Laboratory Press for suggesting the whole project and for his help and encouragement through the process, Jim Watson of Cold Spring Harbor Laboratory for helpful discussions in shaping the idea of the book, Michael Zierler for his expert editing, and John Inglis, Jan Argentine, Carol Brown, Kathy Bubbeo, Maryliz Dickerson, Susan Schaefer, and Denise Weiss of Cold Spring Harbor Laboratory Press for their help in various stages of the project. I also thank Gerald M. Edelman, W. Einar Gall, and the Neurosciences Research Foundation for their support, both intellectual and material, and for having created an environment at The Neurosciences Institute that encourages unconventional thinking. Finally, I thank my wife Dani Grady for her encouragement, support, suggestions, criticisms, and inspiration.

Ralph J. Greenspan
San Diego, California
September 2006

What Are Brains for?

In Mark Twain's 1894 novel *Pudd'nhead Wilson*,[1] the title character is a detective who figures out that the eldest son (and heir apparent) of a large Southern plantation is really the child of one of the slaves. The real heir and the slave child, indistinguishable from each other by skin color due to ubiquitous interbreeding between male slave owners and female slaves, had been switched at birth. "Pudd'nhead" was so named because no one ever thought that he could do much, just as children of slaves were assumed to be intellectually inferior. Twain's switched-at-birth scenario was designed to illustrate what we now refer to as the role of "nurture" (as opposed to "nature") in the shaping of a person, and to skewer the prejudice of classifying slaves (and some detectives) as inferior beings. With today's neurobiological perspective, we would cite these as examples of the brain's ability and propensity to adapt itself and its bearer to the world around it. The child raised as the plantation owner's firstborn showed all of the traits and foibles to be expected of his station, whereas the child raised as a slave showed the corresponding characteristics.

So why do we need a brain? It is the only way that we, or any other creature that moves through the world, can deal with the constantly changing panorama of sights, sounds, smells, tastes, and touches, not to mention gravity, and (for some animals) electrical and magnetic fields. Brains allow us to perceive the world, respond to it, move through it, and act on it. The amount of brain we have, measured as the number of nerve cells (neurons), determines how much of a repertoire for perception, response, movement, and behavior we have at our disposal: relatively little for a single-cell protozoan, somewhat more for a jellyfish of 1,000 neurons, a fair amount for a fruit fly of 100,000 neurons, and quite a lot for a human of 100 million neurons. When it comes to brains, size unquestionably matters.

But aside from size, are all brains fundamentally the same? Even identical twins, whose brains are the same size and genotype, do not really have identical brains. And if that is so, then how likely is it that the brain of a human and that of a jellyfish have anything in common? As it turns out, all brains have much in common. Even

---

[1]Twain M. 1894. *Pudd'nhead Wilson: A tale*. Chatto & Windus, London.

the single-cell protozoan *Paramecium*, which has been likened to a neuron that swims, has much in common with our own nerve cells. Neurons are for signaling, electrical signaling to be precise, and the nature of those signals appears to have evolved very early—before multicellularity—and to have been well preserved ever since, right down to the molecules that make it happen.

Senses go back a long way as well. Chemical sensing is almost certainly the original sense, given that life arose in the liquid environment of the sea and that even bacteria have a non-neuronal version of it. But with the arrival of multicellular animals, separate sense organs arose and with them the ability to see and hear as well as taste, smell, and touch. The mechanisms that make these sensory detection events possible are also well conserved across distant phylogenies, with some variations. Present-day descendants of these primitive detectors and their molecular machinery are found in simple aquatic organisms such as the barnacle and the horseshoe crab *Limulus*.

But senses by themselves do not require a brain, since plants too can "sense" and respond to light, gravity, and humidity. Movement is the sine qua non of animal life: deliberate, directed, internally generated, often rapid movement. Motor systems depend not only on the electrical signals conducted along neurons, but also on the coordination of activity among sets of neurons and muscles. The rapid communication among cells occurs across synapses, which are the specializations at the ends of neurons for chemical signaling. Just as electrically conducted signals are conserved from *Paramecium* to humans, so the chemically transmitted signals between neurons are conserved through evolution. In the context of the neuronal circuitry of a simple brain, these properties can generate patterns of muscle activation that can propel an animal through its environment. Again, we find present-day examples of what likely were primitive, rhythmically coordinated motor systems in the rhythmic contractions of a jellyfish or the undulations of a swimming leech.

Is that it? Are brains nothing more than glorified sensory input and motor output devices? B.F. Skinner, whose viewpoint (known as "behaviorism") dominated American psychology for most of the 20th century, pretty much believed that. Skinner may have also added to this the ability to make associations, as in the conditioning of Pavlov's dogs (who learned to anticipate mealtime after associating the ringing of a bell with the imminent arrival of lunch). "Associative conditioning" also appears to have a deep, evolutionary ancestry and to be a property of very simple brains. Even a simple animal must adapt to its immediate environment. In fact, the ability to modify its signaling can be a property of any synapse, as seen in the sea snail *Aplysia,* and once again, the molecular machinery necessary for these events shows great similarity in all animals.

Skinner theorized that these simple capabilities, associative conditioning in particular, account for all of our behavior, including our mental experience. This represents the most radical, antideterminist view of human nature. Its counterpart was originally proposed by two of the founders of modern genetics and eugenics in the late 19th and early 20th centuries, Francis Galton in England and Charles Davenport in America. Their opposing viewpoint states that human nature and behavior are biologically determined through and through. Mark Twain would certainly have more sympathy for Skinner than for Galton and Davenport, as the story of Pudd'nhead Wilson and the slave versus master upbringing shows. But he was too shrewd an observer of the human condition to go along with Skinner's complete denial of an intrinsic human nature. In fact, Twain states, "With all respect for those ancient Israelites, I can not overlook the fact that they were not always virtuous enough to withstand the seductions of a golden calf. Human nature has not changed much since then."[2]

Whether brains are determined or changeable, they are essential for the richness and diversity of behavior that animals display. How the brain organizes a repertoire of behaviors has been explored in the fruit fly *Drosophila melanogaster*, champion of genetic studies in the 20th century. The range of biological problems that can be answered by studying *Drosophila* is continually expanding. These little flies have shown us not only how similar the genes of all animals are to one another, but also how similar many overt animal behaviors are. The similarities are not just skin- or cuticle-deep, as the case may be, but extend to the underlying mechanisms as well. Circadian rhythms and associative conditioning are prime examples, with findings on sleep and arousal, the sequencing of motor patterns, and perceptual events catching up quickly. And they do this with a relatively modest brain of roughly 100,000 neurons. If one looks to the honeybee, a cousin of *Drosophila* with roughly 1 million neurons, one finds a highly sophisticated ability to make abstract discriminations (e.g., to recognize asymmetry per se) and to find their place in a region as if referring to a map (see Chapter 8).

These postcards from natural history tell us that brains are more than just input–output devices and more than just associatively conditionable, Skinnerian machines. They are complex organ systems, capable of arousal, perception, context-appropriate reactions, anticipation, and sophisticated discriminations; all elements that, in the more complex, neuron-rich environment of the human brain, go together to produce consciousness, feelings, and ideas. We know a lot about the molecular parts that comprise this organ system, a lot about the physiology of its cells, and something about the neuronal circuitry for certain tasks. We have general outlines about how some of the more sophisticated feats are achieved, but few of the steps are known, especially in understanding links among the levels of molecules, cells, and circuits. There remains much that is mysterious. As a prominent neuroscientist has said, "The kidneys just make piss but brains make epistemology."[3] Understanding the brain and the way it performs its functions stretches the limits of our scientific and linguistic abilities.

---

[2] Twain M. 1869. *The innocents abroad: or, The new Pilgrims' progress*, Chapter XLVI, p. 480. American Publishing Co., Hartford, Connecticut.
[3] *BioEssays* **26**: 326–335.

Aside from the intellectual challenge of understanding brains in general, we have some more immediate concerns that relate to understanding brains. Much of our time as human beings is spent trying to understand one another: What did she say? What did he mean by that? Why did they do that? How can we get them to change? These are the grist for our daily conversations and are often of more than idle importance to us. The underlying plot of these conversations is the effect that the goings-on in one person's brain has on those in another person's brain. So in a sense, we spend much of our lives trying to understand one anothers' brains. Granted, we do not always do such a good job of it, but we try nonetheless. This is one justification for trying to understand how brains give rise to behavior in animals. And because none of us wants to submit to being experimented upon, beyond maybe earning $20 to play a video game for a psychology experiment as students, we study animals.

Why study invertebrate brains, as opposed to those of other mammals, if we want to understand our own brains? Evolution has produced a wide range of variations in how animals accomplish tasks and solve problems. Invertebrates include a much greater share of this variation than vertebrates and much more than mammals. As a result, there are some experiments that are only possible to do in certain invertebrates. The examples in this book are drawn from such experiments. At the same time, for all of evolution's variety, there are common threads that run through the workings of brains in all animals. Often, these are more easily revealed in the experimental setups possible with invertebrates and later confirmed in mammals.

There is also a broader reason for wanting to understand the range of possibilities open to brains. Evolution may be seen as one vast experiment in how many ways there are for living things to get by on this earth. No single one of them is necessarily "the best," and all are only as good as their ability to hang on and avoid extinction. Yet the range is not infinite. Not only are all organisms limited in what they can do, but there may be limits to how behavior can evolve based on the limits to what brains can do. Being made of cells that communicate with one another by means of electrical and chemical signals may carry its own inherent limitations. By exploring the boundaries of these limitations in the living world, we may arrive at a deeper understanding of the principles that govern the architecture and workings of nervous systems.

CHAPTER 1

Avoidance and the Single Cell
Ionic Signals

Paramecium at times accepts things that are useless or harmful to it, but perhaps on the whole less often than does man.

H.S. Jennings

Paramecium, the freshwater protozoan that resembles the pattern elements in paisley fabric, has a long history as a subject of study. Beginning with van Leeuwenhoek (Fig. 1.1) in the 17th century, who first identified them in preparations of pond water ("infusaria") using his newly invented microscope, they became subjects for the study of behavior in the early 20th century when one of the leading experimental biologists of the day, H.S. Jennings, showed that these little unicellular

Figure 1.1. Antony van Leeuwenhoek (1632–1723), Dutch inventor of the microscope, who first observed *Paramecium* in preparations of pond water.

creatures had a surprisingly rich behavioral repertoire. Swimming by means of the many thousand cilia covering their single-cell bodies, they would speed up if bumped from behind or back up and swim off in a new direction if bumped into from the front, and respond to a whole variety of stimuli. Jennings was so impressed with their capabilities that he thought they might actually be conscious.

Seventy years later, their behaviors were shown to be the result of electrical signals coursing through them, signals that are virtually identical to those that course through the neurons of animals with full-blown nervous systems. Whether *Paramecium* actually share consciousness with their more complex animal relatives is dubious, but there is no question that they do share in the major currency of communication used by complex nervous systems: electrical currents carried by the charged ions (e.g., Na^+, Ca^{2+}, and K^+) found in the fluid inside and outside of cells.

Electrophysiological Control of Swimming Behavior

Paramecia live in ponds, streams, and stagnant pools where they swim, consume bacteria, avoid predators, and occasionally mate. This short list of activities encompasses the vast majority of animal behavior and therefore most of what nervous systems have evolved to perform. These creatures navigate through the water by sensing when they bump into something, or when something bumps into them, and also by detecting changes in the chemical composition and temperature of the water (Fig. 1.2).

The neural strategy for *Paramecium*'s navigation works by converting a mechanical stimulus—being bumped—into a local electrical signal that is generated by the movement of charged ions through the cell's membrane and is confined to the stimulated region of the cell. This local electrical signal then triggers a cell-wide electrical response that alters the beating of their cilia, and thus alters their swimming. As these electrical responses subside, their swimming returns to normal. Electrical signaling of this sort has the virtue of being very fast and capable of spreading over large cellular

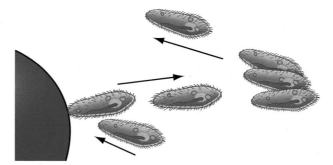

Figure 1.2. Avoidance response of *Paramecium*. When they bump into an obstacle, they reverse their swimming direction, turn, and head off in a new direction.

distances. It appears to be a eukaryotic invention that may have been required for the large cellular sizes that eukaryotic cells attained when they broke out of their prokaryotic constraints.

The electrical signals in *Paramecium* (and in all cells) are generated by the movement of ions across the cell's membrane and make use of the fact that there is a charge difference (an electrical potential) across the cell membrane. The inside of the cell is more negatively charged than the outside, producing a voltage difference of approximately −30 mV in a normal pond-water-like medium. The negative voltage arises from the uneven distribution of certain ions; K^+ concentration is higher inside and Ca^{2+} and Na^+ concentrations are higher outside. This uneven ion distribution is produced, in part, by the action of particular proteins embedded in the cell membrane, known as transporters, that actively pump ions such as Ca^{2+} and Na^+ out of the cell's cytoplasm. Another source of the charge difference is the presence of an excess of negatively charged groups on the many proteins inside the cell (which is true of all cells). None of these differences in charge distribution would make any difference if the cell membrane itself were not predominantly composed of hydrophobic lipids. Charged ions will not penetrate the lipid portion of membrane and so it serves as an insulating material.

The cell generates an electrical signal when another type of protein in its membrane, known as an ion channel, is induced to open (see next section and Figs. 1.12 and 1.13 below). These proteins create a pore through the hydrophobic membrane that permits ions to pass through, for example, Na^+, K^+, or Ca^{2+}, depending on the particular type of channel. Because Na^+ has a higher concentration outside the cell, it will rush in, creating an inward electrical current and causing the charge difference (voltage) across the membrane to become more positive from outside to inside. Because K^+ has a higher concentration inside the cell, it will flow out, creating an outward electrical current and causing the voltage to become more negative. Negative ions, such as Cl^-, flowing into the cell also cause the voltage to become more negative.

At this point, the insulating properties of the lipid membrane come into play to prevent the displaced ions from immediately escaping back to their original location. This allows the flux of ions, the "signal," to spread inside the cell without being instantly dissipated. The electrical circuit is then completed by the sporadic escape of ions out of the cell. Eventually, a normal balance is restored when the transporter pumps expel the newly entered excess ions or import the recently lost ions. Now back to *Paramecium* swimming.

Local Signals

When a *Paramecium* is bumped near its back end, the cell membrane is mechanically deformed, activating a particular set of mechanosensitive ion channels. These channels are induced to open, which results in a rapid flow of K^+ across the membrane (known as a K^+ current). Because the concentration of K^+ is higher inside than

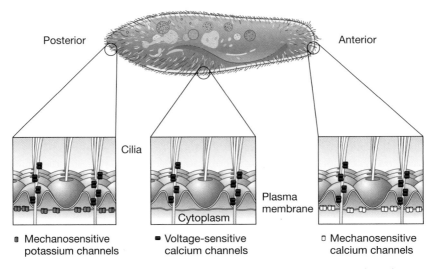

Figure 1.3. Membrane specializations in *Paramecium*, showing localization of mechanosensitive potassium channels in the posterior plasma membrane, voltage-sensitive calcium channels in the extensions of the plasma membrane around each cilium (which are all over the cell), and mechanosensitive calcium channels in the anterior plasma membrane.

outside the cell, the flow is outward, the cell loses positive charges, and the voltage across its membrane becomes more negative (i.e., it becomes hyperpolarized).

When the membrane potential becomes hyperpolarized, the cilia beat faster and the cell shoots forward. The mechanosensing potassium channels are localized near the posterior end of the cell (Fig. 1.3). The change in membrane potential is largest where the deformation occurs and then decreases as it spreads along the membrane from this point. For this reason, the change in ciliary beating is strongest at the back end of the cell. The size of this change in membrane potential is proportional to the strength of the bump, increasing with the bump's intensity. (Membrane potential is measured by immobilizing the cells in gelatin and inserting a very fine, hollow glass needle [microelectrode] through the membrane. The microelectrode is filled with saline solution and connected to a wire so that the voltage difference between the inside of the cell and the liquid medium around it can be measured with an extremely sensitive voltmeter. The "traces" shown in subsequent figures in this book represent records of the voltage difference between the inside and outside of a cell recorded on a millisecond timescale, fast enough to detect the very rapid signals that these cells generate.)

If *Paramecium* encounters an obstacle and the bump occurs at the front end of the cell, another class of ion channels is activated to open up, allowing a mixture of cations to enter, a major one being Ca^{2+}. Because these are positively charged

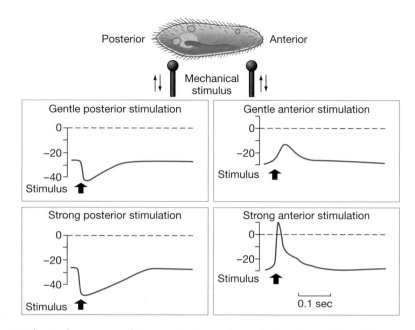

Figure 1.4. Electrical responses of *Paramecium* to mechanical stimulation at its anterior or posterior ends. Traces (*blue*) represent changes in membrane potential in millivolts (ordinate) versus time (abscissa) after gentle (*upper graphs*) or stronger (*lower graphs*) stimulation. In each graph, the *dashed line* represents zero voltage across the membrane and the *arrows* represent the timing of the stimulus. The timescale is shown at *lower right*. The membrane responses are graded in response to graded stimuli and are hyperpolarizing at the back end (*left graphs*) and depolarizing in the front end (*right graphs*).

ions, they drive the membrane potential in the positive direction (i.e., it depolarizes), and like the back-end bump described above, the magnitude of membrane depolarization is strongest near the site of the stimulus and decreases as it spreads from there (Fig. 1.4). And also like the back-end bump, the size of the change in membrane potential is proportional to the strength of the bumping stimulus.

The cell uses different ion channels for each of its responses. The determination of which ion is responsible in a given case is made by changing the external concentration of each ion in a stepwise manner and measuring the response at each step (Fig. 1.5). When the critical ion's concentration is altered, the response changes.

Global Signals

What makes the front-end bump special is that when the membrane potential becomes sufficiently positive, it triggers a second, larger response that spreads instantly throughout the cell. This large response differs from the graded responses

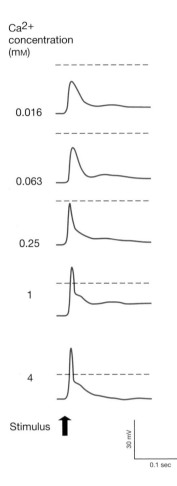

Ca2+
concentration
(mM)

0.016

0.063

0.25

1

4

Stimulus

30 mV

0.1 sec

Figure 1.5. Response to mechanical stimulation of the anterior surface as a function of extracellular $[Ca^{2+}]$. Traces (*blue*) show the change in membrane potential after mechanical stimulation. Timing of stimulation is indicated by the *arrow* at the bottom of the column. Numbers to the right are increasing concentrations of Ca^{2+} (mM), with other ions held constant. The *dashed horizontal line* in each trace represents zero membrane potential under those conditions. Shown at *lower right* are scale bars for membrane potential (mV; ordinate) and time (sec; abscissa). Calcium dependence of the response is shown by the increase in the size of the response with increasing extracellular Ca^{2+} concentration.

to mechanical stimulation in the sense that once triggered, it results in a stereotypical electrical signal: The membrane potential rapidly rises to a peak positive value, a peak that is always the same. This phenomenon, known as an "action potential," is widely found in neurons and muscles, as well as in *Paramecium* (and probably many of its relatives). Action potentials are found wherever long distances must be spanned (as in a large neuron) or whenever a rapid and uniform change in membrane potential in a large cell (as in a *Paramecium* or a muscle) is needed. They are produced by a type of ion channel that opens in response to a positive voltage change (depolarization). These are known as voltage-sensitive ion channels and they ensure the rapid spread of the signal. As one channel opens and depolarizes its patch of membrane, the inflowing ions spread around this patch inside the cell and trigger the opening of neighboring channels. These, in turn, depolarize their neighborhoods and trigger the opening of still more channels nearby. The process starts with mechanosensitive depolarization at the front end of the cell and ends when the entire cell is depolarized. This process occurs nearly instanta-neously. (Action potentials are so rapid and bullet-like in their timing that cells are often described as "firing" action potentials, which are often called "spikes" based on the shape of their voltage traces.)

The stereotypical peak of the action potential is determined by the concentra-tion of the particular ion used by the voltage-sensitive channel. In *Paramecium*, the action potential is produced by voltage-sensitive Ca^{2+} channels, and thus it is the Ca^{2+} concentration difference between the inside and outside of the cell that sets the peak. When enough Ca^{2+} enters the cell, the inside concentration equals the outside concentration and the membrane depolarization can go no further.

How does depolarization of the whole *Paramecium* cell alter its swimming behavior? The behavioral response consists of a reversal of the beating direction of all the cilia, brief backward swimming, swiveling, and then heading off in a new direction. The driver for ciliary reversal is the same as the driver of the action potential: Ca^{2+} ions. The concentration of free Ca^{2+} inside the cell is normally very low, <0.1 μM. When intracellular calcium reaches a certain concentration (>0.1 μM) in the vicinity of the cilia, their usual posterior power stroke reverses so that they beat forward and the direction of swimming thus becomes backward (Fig. 1.6). As Ca^{2+} is pumped out, ciliary beating reverts to its original orientation. Fig-ure 1.7 summarizes the steps of the avoidance reaction.

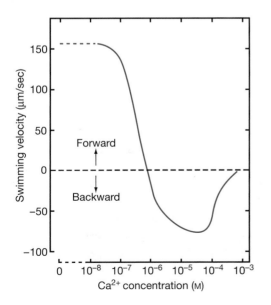

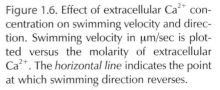

Figure 1.6. Effect of extracellular Ca^{2+} con-centration on swimming velocity and direc-tion. Swimming velocity in μm/sec is plot-ted versus the molarity of extracellular Ca^{2+}. The *horizontal line* indicates the point at which swimming direction reverses.

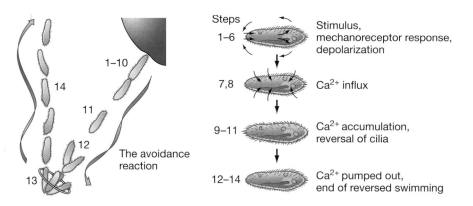

Figure 1.7. Steps of the avoidance reaction: (1) forward swimming until an obstacle is encountered; (2) deformation of anterior membrane upon collision with obstacle; (3) local opening of mechanosensitive channels; (4) inward Ca^{2+} current through these channels in the stimulated membrane; (5) passive spread of this current caused by insulating properties of the membrane; (6) depolarization of cell membrane, producing (7) opening of voltage-sensitive Ca^{2+} channels; (8) Ca^{2+} action potential; (9) rise in intracellular Ca^{2+} concentration; (10) cilia reverse beat; (11) cell swims backward; (12) Ca^{2+} pumped out; (13) intracellular concentration of Ca^{2+} drops and cilia resume normal orientation; and (14) cell swims forward.

Mutant strains of *Paramecium* that lack the action potentials also lack the avoidance response (Fig. 1.8). These have been dubbed *"pawn"* because they can only move forward, like the chess piece.

Critical to the swimming responses of *Paramecium* is the localization of the various channel types to different regions of the cell's membrane. Mechanosensitive potassium channels are in the posterior end, whereas mechanosensitive calcium channels are in the anterior membrane. This is seen in the differential response to prodding at different points along the cell's membrane (Fig. 1.9). The voltage-sensitive calcium channels that produce action potentials are localized in patches of

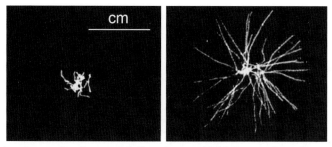

Figure 1.8. Swimming tracks of normal (*left*) and *pawn* (*right*) *Paramecium* after exposure to Ba^{2+} ions, which mimics the avoidance response.

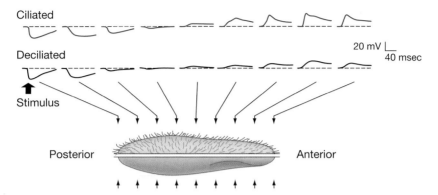

Figure 1.9. Change in the electrical responses to mechanical stimulation along the anterior–posterior axis of *Paramecium*. (*Bottom panel*) Points of stimulation along the length of the cell. (*Upper traces; blue*) Membrane potentials after stimulation in intact cells with cilia; (*lower traces; black*) responses of cells stripped of their cilia. (*Boldface arrow*) Timing of mechanical stimulation relevant to the two traces above it; comparable stimuli are given for each of the other traces. Removing the cilia eliminates the calcium-based action potential (compare the last four *blue* and *black* traces on *right*) while maintaining the graded receptor potentials at each end. (*Right*) Scale bars for membrane potential and time.

membrane covering each cilium, and are thus situated at the crucial spot to regulate the entry of calcium in the immediate vicinity of the cilium. The calcium channels on each cilium are initially activated by depolarization in the vicinity of the anterior mechanosensitive channels, and then depolarization spreads almost instantly throughout the cell as patch after patch of voltage-sensitive calcium channel opens. Cells that have had their cilia stripped off (deciliated) lack action potentials, but retain the local anterior and posterior responses, indicating that channels for the local responses are in the anterior and posterior plasma membrane, whereas the voltage-sensitive calcium channels for the action potential are in the membrane immediately around each cilium (Figs. 1.3 and 1.9).

Sensing the Environment

Bumping into obstacles is not the only (or even the major) hazard a *Paramecium* faces in life. Freshwater environments can be very unstable places, subject to rapid and drastic changes in salt and mineral concentrations as the result of rainfall, drought, and flowing water. It is therefore not surprising that *Paramecium* exhibits avoidance behavior in response to increases in Na^+ or Mg^{2+} ion concentrations. Each of these ions has its own special class of channels, and mutants have been isolated that selectively eliminate the responses.

A mutant called *fast-2* shows no Na^+ avoidance response, continuing to swim happily forward into a concentration of the ion that would stop a normal *Parameci-*

um in its tracks and cause it to reverse. When physiological responses are recorded from these mutants, they are found to be lacking in the ability to permit Na^+ entry. This Na^+ influx requires the presence of Ca^{2+} and is thus called a "Ca^{2+}-dependent Na^+ current" (Fig. 1.10). This is likely to be a sodium channel that is regulated by its binding of Ca^{2+} ions.

When a mutant strain known as *eccentric* is exposed to a high concentration of Mg^{2+}, it fails to reverse its swimming. Recordings made from these mutants indicate that there is no calcium-dependent magnesium current, which is present in normal cells (Fig. 1.10). This Ca^{2+}-dependent Mg^{2+} current is likely due to a magnesium channel that is regulated by its binding of Ca^{2+} ions.

Recovery

All of the electrical responses described above are transient. Without mechanisms for returning to the original (resting) membrane potential, the cell would die. The restoration of the original membrane potential is due to a combination of rapidly acting channels and slower acting pumps. The channels redistribute charge by allowing additional ion movements across the membrane, and the transporter pumps restore the initial intracellular concentrations of ions.

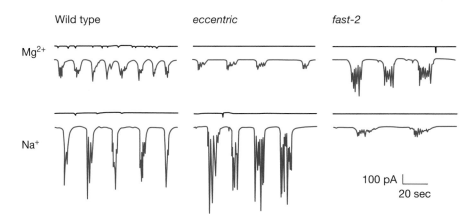

Figure 1.10. Membrane responses in normal, *eccentric*, and *fast-2 Paramecium* in the presence of high extracellular Mg^{2+} (*upper two traces*) or high extracellular Na^+ (*lower two traces*). These responses illustrate the physiological defects underlying the behavioral differences among strains. Normal (wild type, *left*) *Paramecium* shows robust changes in membrane potential in response to both kinds of ions. *eccentric* mutants (*center*) respond normally to Na^+ but not to Mg^{2+}, and *fast-2* mutants (*right*) respond to Mg^{2+} but not to Na^+. Shown are responses before (*black*) and after (*blue*) presentation of the ions. These traces are displayed oppositely from those shown previously, so that what corresponds to a depolarization goes downward. They record membrane current (in picoamps) instead of voltage. (*Lower right*) Scale bars for membrane current and time.

The depolarization induced by the inward calcium current is initially counter-acted by a set of potassium channels that are triggered to open by the depolarization. These voltage-sensitive K^+ channels short-circuit the voltage change by repolarizing the membrane as K^+ flows out of the cell. Moreover, the calcium channels them-selves stay open for only a short time, because they are induced to close by the influx of Ca^{2+} in the cell (they actually bind Ca^{2+} ions on their intracellular side). Reinforc-ing these processes is another set of potassium channels that open more slowly in direct response to the buildup of Ca^{2+} in the cell. And finally, a different kind of restorative mechanism is provided by membrane transporter proteins that use ATP energy to eliminate free Ca^{2+} from the cytoplasm by pumping it either out of the cell or into internal membrane compartments.

Degenerate Mechanisms

Why should *Paramecium* have so many separate mechanisms for accomplishing the same end, that is, returning to their original steady state? Questions of "Why?" are always difficult to ask with respect to evolution, because the answer is always "Because that is the way it happened." But there is something evolutionarily important to say on the subject. Multiple, nonidentical mechanisms for accomplishing the same end, even within the same organism or cell, are widespread in the living world and are likely to be essential elements of evolutionary fitness and flexibility. This is not the same as redundancy. Redundancy occurs when there are identical copies or versions of the mechanism, such as duplicated genes or duplicated circuits. Instead, the more accurate term is "degenerate," as in the concept of degenerate orbitals in the structure of an atom, which have the same energy but different shapes, or in the genetic code in which different triplet base pairs encode the same amino acid. Thus, *Paramecium* has degen-erate mechanisms for returning its membrane potential to the resting state.

The Molecular Basis for Ion Currents

The idea of ion channels originated from electrophysiological studies in the 1950s similar to those described above, in which currents carried by specific ions were identified and each current had a characteristic stimulus and time course for starting and stopping. In the ensuing half century, many of these ion channels were identi-fied as transmembrane proteins, and in a few cases, structure/function studies showed how they open and how they act as selective filters.

Potassium channels are the best understood class of ion channels, in large mea-sure because they are also among the most ancient of ion channels. Aside from being virtually ubiquitous among eukaryotes, including those that lack nervous systems altogether, gene sequences for potassium channels have also been found in many species of bacteria and archaea (Fig. 1.11). In the extreme, a K^+ channel gene has

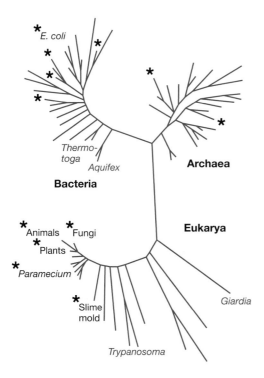

Figure 1.11. K$^+$ channels found (thus far) among prokaryotes. The display is a "universal phylogenetic tree" that positions the various species according to the extent of similarity in the DNA sequence of a representative gene, their single-stranded ribosomal RNA molecules. Animals, plants, and fungi constitute only a small portion of the total biological diversity, even among the eukaryotes (eukarya). *Giardia*, the parasite well known to backwoods hikers, is one of the most primitive eukaryotes. *Thermotoga* and *Aquifex* are two of the most primitive bacteria. *Asterisks* denote that K$^+$ channels have been identified.

been found in the genome of a virus that infects the unicellular green alga *Chlorella*, which, in turn, lives as an endosymbiont in certain species of *Paramecium*. *Paramecium* itself has a whopping 298 distinct sequences for K$^+$ channel genes (a record among organisms surveyed and three times the number found in humans). The importance of phylogenetic ubiquity to our understanding of K$^+$ channels can be traced to the fact that the initial structural studies responsible for our detailed picture of their mechanism were made possible by the large amounts of pure channel protein obtainable from several bacteria and archaea, including the relatively simple K$^+$ channel protein from the bacterium *Streptomyces lividans*. With this structure as a starting point, studies of neuronal potassium channels followed.

Structural studies have shown that, when assembled, the four K$^+$ channel subunits form a funnel-shaped pore through the membrane. This shape is critical for the way in which they achieve their selectivity for K$^+$ ions. K$^+$ ions normally attract a shell of water molecules around them. The K$^+$ channel permits the entry of this hydrated shell into its wide extracellular end and then coaxes the ion to shed its water and enter the narrow pore, whose size is snugly fit for K$^+$ (Fig. 1.12). Na$^+$ and Ca^{2+} ions are excluded because the K$^+$ channel's configuration does not overcome the greater energy required to shed their water molecules.

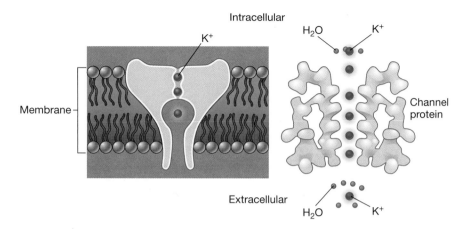

Figure 1.12. K^+ ions are shown exiting the cell through a voltage-sensitive potassium channel: (*left*) embedded in the membrane lipid bilayer with a wide opening to allow entry of potassium ions (*blue*) from the intracellular side, and (*right*) showing a procession of potassium ions (*larger blue beads*) initially surrounded by water molecules (*smaller blue beads*), then shedding them as the ion enters the pore and becoming rehydrated as it exits in the extracellular space.

Because all K^+ channels must filter K^+ ions, the pore-forming domain is the one that is shared by all of them (Fig. 1.13). But there are many different ways to open a K^+ channel: by voltage, by the binding of ligand (an ion such as Ca^{2+} or a small molecule such as cyclic guanosine monophosphate [GMP]), and by mechanical distortion of the membrane. The structural events associated with these gating mechanisms involve conformational changes induced by the particular stimulus—voltage, ligand, or mechanical stress. Other kinds of ion-selective channels appear to work by analogous mechanisms.

Mechanosensitive channels are also found all over the phylogenetic tree (Fig. 1.14). Unlike K^+ channels, however, the similarities across the three major domains of life are not all based on shared sequences. Various organisms within bacteria, archaea, and eukarya contain membrane proteins that open up to let ions pass upon mechanical distortion of the membrane. One major class of these channels is exclusively eukaryotic (so far), sharing sequence similiarity among worms, flies, mammals, and fungi. The other major class shares sequence similarity among bacteria, archaea, fungi, and plants.

Studies on purified mechanosensitive channels from the bacterium *Mycobacterium tuberculosis* have shown that the pentameric protein opens when tension is applied asymmetrically to the membrane. Symmetrical pressure has little effect, but one-sided pressure from either the inside or the outside distorts the lipid bilayer and the channels respond by opening (Fig. 1.15).

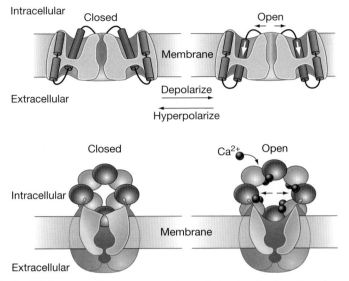

Figure 1.13. (*Top*) Voltage-sensitive potassium channel showing its closed configuration when the membrane is hyperpolarized, and its open configuration after depolarization. Channel is shown outlined in *blue,* with *black* and *blue cylinders* representing portions of it. (*Blue cylinders*) The voltage-sensitive segments of the channel that move toward the outward side of the membrane (*arrows*) when the membrane is depolarized. (*Bottom*) Calcium-activated potassium channel showing closed configuration in the absence of calcium and open configuration after binding Ca^{2+}. (*Blue shapes* and *large gray ovals*) The channel; (*small gray beads; right*) Ca^{2+} ions.

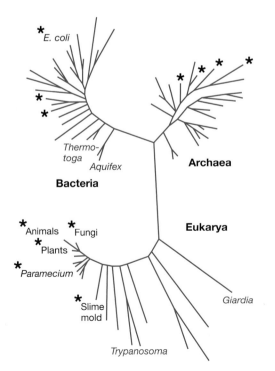

Figure 1.14. Range of species in which mechanoreceptors have been found (*asterisks*). Placement on the universal tree of life (see Fig. 1.11) of species in which mechanoreceptors have been identified and their DNA sequences compared.

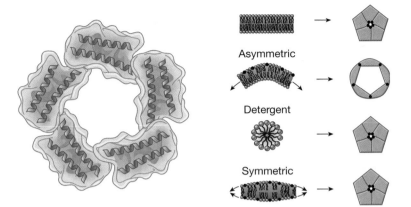

Figure 1.15. Structure of the mechanosensitive channel from the bacterium *Myobacterium tuberculosis* (*left*), viewed as if looking at the outside face of the membrane. Channel configuration (*far-right column*) is shown under various conditions of membrane distortion (*middle column, top to bottom*): no distortion, asymmetric distortion, detergent (no distortion), and symmetric distortion. Only under asymmetric distortion does the channel open.

Evolutionary Origins of *Paramecium's* Neural Strategy of Behavior

Why should unicellular organisms harbor the same repertoire of ion channels as their metazoan cousins? Leaving aside "Because that is the way it is!," a plausible explanation is that the channel genes originally evolved to allow cells to adapt to changing osmotic conditions, because osmotic variations (rain versus shine) occurred on our planet even in its infancy. The mechanosensitive channels in bacteria seem to be required for such responses: *Escherichia coli* that are doubly mutant for both of their mechanosensitive channels will lyse (explode) after even the mildest change in solute concentration. Potassium channels may have originally served a similar function, and vestiges of this history may still be found in a class of potassium channels present in virtually all cells that provide a constitutive low level of ionic flux known as a "leakage" current. These are important for completing ion current circuits (see above).

Under the influences of time, genetic changes (such as the duplication of individual genes and their subsequent divergence in sequence to acquire new sites and domains), and selection pressures, a more versatile set of channels could have emerged with different ion filters and diverse modes of regulation, such as voltage sensitivity, Ca^{2+} dependence, and self-inactivation. Once such genes came on the scene, further selection could have resulted in their being organized and localized into the apparently seamless ensemble of components that regulate swimming behavior.

In comparison to animals with nervous systems, the repertoire of *Paramecium's* behavior is not large. But considering that it is only a single "neuron," it does not do badly. Its neural strategy combines in one cell a number of tasks that become subdivided among many cells in a nervous system. These include the presence of special-ized structures that respond in a graded fashion to sensory stimuli, the integration of local and general signaling mechanisms, the ability to respond to sensory input with appropriate changes in motor output, and a recovery mechanism to return to pre-stimulus behavior.

Why would this kind of electrical mechanism have evolved for a single cell? What was wrong with plain old flagellar beating modified by diffusable chemical sig-nals as in other unicellular eukaryotes, or protein–protein interactions as in bacteria? Electrical signaling has the advantage of being able to cover large distances very rapidly, much more quickly than chemical diffusion, and *Paramecium*, at 180 μm, is considerably larger than a bacterium (1 μm) or a choanoflagellate (10 μm), the near-est unicellular relative to animals.

The uncanny resemblance of the electrical signaling mechanisms in *Paramecium* to those of animal neurons raises the question of the ciliate's place in phylogeny. Are their ancestors unicellular protists, or are they unicells descended from multicellular ancestors, as has been suggested to be the case for yeast based on comparisons of DNA sequences? Such comparisons place *Paramecium* deeply in a neighborhood of eukaryotic unicells with no metazoan cousins on any nearby branches (Fig. 1.16). This argues that their neuron-like properties were present in the unicellular eukary-ote that was ancestral to *Paramecium* and also to animals, a suggestion that is sup-ported by the presence of ion channels in prokaryotes (see above).

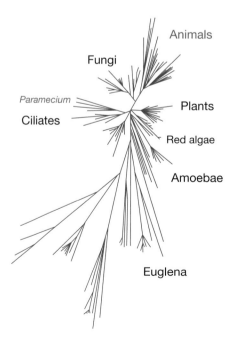

Figure 1.16. Phylogenetic tree of eukaryotes showing relationships among animals, plants, fungi, and various unicellular protozoans based on several representative genes (actin, tubulin, EF-2). *Paramecium* is shown among the ciliates. *Paramecium* and animals cluster far from one another.

Perhaps there are many other protists or even prokaryotes with equally sophisticated and neuron-like electrical signaling systems that have just not been studied yet. Given that the biological diversity of the unicellular world dwarfs that of metazoans, there are likely to be many surprises awaiting the energetic and the curious among us.

Further Reading

Eckert R. 1972. Bioelectric control of ciliary activity. *Science* **176:** 473–481.

Jennings H.S. 1906. *Behavior of the lower organisms*. Columbia University Biological Series, The Macmillan Co., New York.

Kung C. and Blount P. 2004. Channels in microbes: So many holes to fill. *Mol. Microbiol.* **53:** 373–380.

Martin A.R., Wallace B.G., Fuchs P.A., and Nicholls J.G. 2001. *From neuron to brain*, 4th edition. Sinauer Associates, Sunderland, Massachusetts.

Saimi Y. and Kung C. 1987. Behavioral genetics of *Paramecium. Annu. Rev. Genet.* **21:** 47–65.

CHAPTER 2

You Can't Run but Maybe You Can Hide
Chemical Signals

In sessile Cirripedes [barnacles], vision seems confined to the perception of the shadow of an object passing between them and the light; they instantly perceived a hand passed quickly at the distance of several feet between a candle and the basin in which they were placed.

Charles Darwin

Barnacles (Fig. 2.1) are famous for their sedentary lifestyle. As they approach adulthood, they attach to a convenient rock or ship bottom with a tenacity that puts superglue to shame. From then on, they spend the rest of their lives filtering the water that flows or splashes by them for plankton and other microbial treats. As a means of protecting their internal organs, which are exposed while feeding, they quickly close their shells if there is a sudden decrease in illumination (Fig. 2.2)—the "shadow reflex" described by Darwin. A passing shadow is a pretty good indicator of a predator, and if there is no predator, they are not set back by much.

Darwin focused on these creatures at a crucial time in his life. As a prelude to his magnum opus, *On the Origin of Species*, he took an 8-year detour into the natural history and morphological classification of barnacles. Why did he choose to do this at the very moment at which he was ready to reveal his theory of evolution to the world? Perhaps it was displacement activity arising from his fear of coming out with a theory that he knew would get him into a lot of trouble. Perhaps he was anticipating the criticism that he had no credentials as a biologist. His training was in geology, and even though he had been the official naturalist for the H.M.S. *Beagle*, sending samples back to England, in the scientific circles he needed to conquer, he had

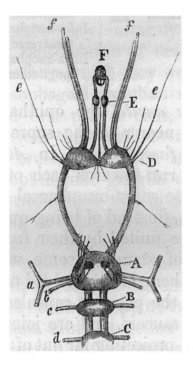

Figure 2.1. Central nervous system of the barnacle *Balanus tintinnabulum* from Darwin's treatise on barnacles.

no credentials as a zoologist. So from 1846 to 1854, he immersed himself in barnacles. At one point, he wrote to a friend, "I hate a Barnacle as no man ever did before, not even a sailor in a slow-sailing ship."[1]

Responding to Light

The barnacle has several simple eyes that are exposed to light when its shell is open. Each one is a cluster of a few photoreceptor cells with no lens, which means that no image is formed. This is a sensory organ that signals "yes," "no," and "how much," rather than "what am I seeing?" Any of these eyes can initiate the shadow reflex. All are exquisitely sensitive to a sudden decrease in illumination, which is what happens when a shadow falls across the barnacle. The photoreceptor cells respond to changes in light intensity with a change in membrane potential. This change is proportional to the light intensity and lasts as long as the stimulus (Fig. 2.3). Increases in light intensity depolarize the photoreceptor cell's membrane. Decreases in light intensity hyper-

---

[1] Charles Darwin's letter to W.D. Fox of October 24, 1852. Letter Number 1489, Darwin Correspondence Online Database Collection, University Library, Cambridge.

Figure 2.2. Shadow reflex of the barnacle *Balanus balanus*, in which its tendrils (cirri) withdraw into its shell (operculum).

polarize it. In addition to being able to respond rapidly, these cells are also capable of responding sensitively over a very wide range of light intensities. This seems appropriate given that there can be a 10,000-fold difference between sunlight and shadow.

The light-induced voltage change is due to the opening of Na^+ channels at the photosensitive end of these cells, and this event comes as the last step in a chain of reactions that begins with the absorption of light by the cell's light-sensitive photopigment (see below). Any change in membrane potential spreads inside the cell, including down the cell's long axon to its terminal (Fig. 2.4). The spreading of this signal is due entirely to the insulating properties of the membrane. There is no mechanism like *Paramecium*'s action potential to regenerate the signal as it goes, and so it

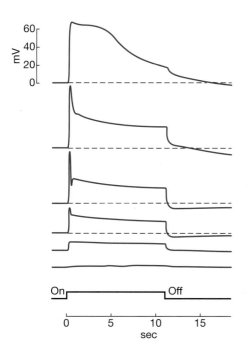

Figure 2.3. Electrical responses of a barnacle photoreceptor to a 10-sec pulse of light. The onset and duration of the light pulse is shown in the bottom trace (step up indicates light "On"), and the response of the cell to pulses of increasing light intensity are shown in ascending order in the middle and upper (*blue*) traces. (*Dashed line*) Resting membrane potential of the cell.

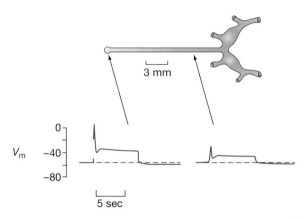

Figure 2.4. Passive spread and decline of the light-induced membrane potential down the barnacle photoreceptor axon from the eye (*left*) to the brain (*right*). Recording sites (*top*) are shown close to (*left*) and far from (*right*) the photoreceptor. The decrement of the signal is seen in the two record-ed traces from each site (*bottom; blue*). Traces represent membrane potential (V_m) versus time at each site.

decays along the way (see comparison of traces recorded from either end of the cell in Fig. 2.4). The insulating properties of the lipid membrane are adequate, however, to ensure that enough of the signal reaches the end of the cell in order to enter the barnacle's little brain.

A Different Kind of Signal

At this point, the mode of communication changes. Animal nervous systems are com-posed of a large number of separate neurons, and electrical signals cannot jump between cells like a static electricity generator. The terminal region of the photore-ceptor axon is formed into a structure, the synapse, that is specialized for sending a chemical signal from one cell to the next (Fig. 2.5). Release of the chemical that trans-mits the signal is triggered by depolarization of the membrane at the axon terminal. From there, the transmitter diffuses the short distance (~10 nm) to the membrane of the adjacent cell (known as the barnacle's "I" cell). Proteins on the I cell membrane are capable of binding the neurotransmitter and opening attached ion channels. As a result, binding of the neurotransmitter by these receptor/ion channels produces a change in membrane potential in the I cell and initiates a new signal in that cell.

In the barnacle's case, the photoreceptor terminal releases the chemical hista-mine. Best known for its role in allergic reactions and for producing runny noses and watery eyes, histamine is also a neurotransmitter. In the barnacle, histamine recep-tors on the postsynaptic I cell respond by opening Cl⁻ channels. Because chloride is

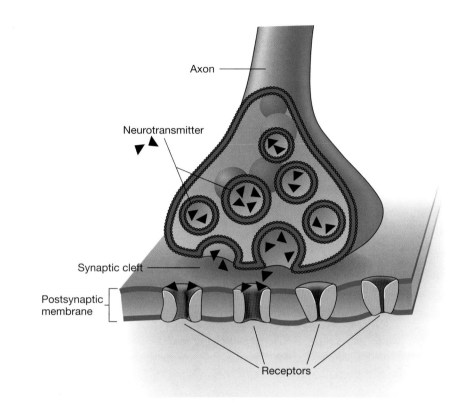

Figure 2.5. Nerve terminal showing vesicles full of neurotransmitter (*triangles*) and receptors (*"donuts"*) on the postsynaptic membrane. When the nerve terminal is depolarized, vesicles fuse with the membrane and release transmitter, which then binds to receptors on the postsynaptic membrane. The receptors (in this case) open ion channels through the membrane.

more abundant outside than inside the cell, the opening of these channels causes negative ions to flow into the cell and hyperpolarize it (Fig. 2.6). When the I cell hyperpolarizes, its own terminal releases less transmitter. As a result, "lights on" inhibits (hyperpolarizes) the I cell and "lights off" (as in a shadow passing over) depolarizes it. In other words, these cells are poised to set off an alarm by causing rapid and intense signaling activity in the barnacle's brain at the moment that light intensity decreases. The sensitivity of this response is so great that a stepwise reduction in light intensity triggers responses in the postsynaptic cell at each descending step (Fig. 2.6). These responses, in turn, travel down the axon of the I cell and induce transmitter release at its terminal, thus triggering depolarization and action potentials in the next cell down the line, the "A" cell. At the end of this tag-team process, involving several more cells in the brain, muscles are stimulated to contract and the barnacle closes up.

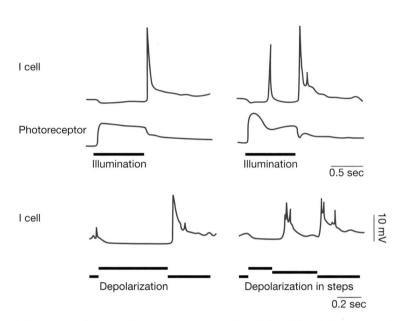

Figure 2.6. Responses of the I cell's membrane potential to light off (*top-left traces*) and to a step down in light intensity (*top-right traces*). Similar responses to direct depolarization of the I cell's postsynaptic membrane (*bottom-left traces*) and to a step down in its membrane potential (*bottom-right traces*). The I cell is postsynaptic to the barnacle photoreceptor. Traces labeled "I cell" are membrane-potential recordings from that cell, and traces labeled "Photoreceptor" are membrane-potential recordings from that cell. The *black* traces under the "Photoreceptor" traces indicate the duration of the light pulse, and the *black* traces at the bottom of the figure indicate the time of steps up and down in membrane potential artificially applied to the I cell. In all instances, the I cell responds when there has been a step down in its membrane potential.

The Strategy of the Shadow Reflex

The shadow reflex is thus a product of an alternating sequence of chemical and electrical signaling events: (1) light absorption (chemical) induces a membrane depolarization (electrical) that (2) travels down the cells' axons to the terminals where it induces release of histamine (chemical), which then opens Cl⁻ channels on the I cells' postsynaptic membrane (electrical). This (3) hyperpolarizes the I cells so that they (4) release less transmitter to their postsynaptic A cells, closer to the brain. When a shadow passes over, the light reaction ceases, the photoreceptors repolarize their axons, their synapses stop transmitting histamine, and, as a result, the postsynaptic I cells depolarize, which stimulates the release of transmitter from their terminals and triggers the A cells to fire action potentials (Fig. 2.7). The A cells then drive a cluster of cells called the motor ganglion that drive neurons contacting the muscles, which (finally!) drive the muscles responsible for closing the barnacle's shell (Fig. 2.8).

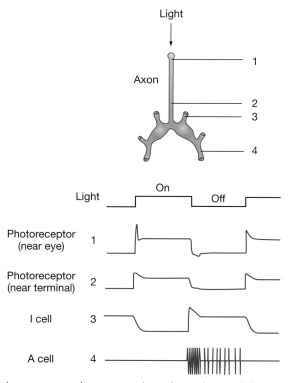

Figure 2.7. Electrical responses at four stages along the pathway of the shadow reflex: (*1*) the response to light absorption by the photoreceptor cell, (*2*) the signal as it travels down the cell's axon, (*3*) the response from the I cell (postsynaptic to the photoreceptor), and (*4*) the response from the A cell (postsynaptic to the I cell). Light on and off is shown by the square steps in the upper (*black*) trace. (*Bottom*) Recordings *1–3* represent recordings of the cells' membrane potentials, and *4* represents action potentials fired by the A cell as recorded near the cell.

Figure 2.8. *Balanus* operculum shown open (*left*) and closed (*right*).

Nervous systems are thought to have evolved to enable multicellular creatures to move around. Usually, this involves chasing or being chased, and the speed of electrical signaling makes quick responses possible. For adult barnacles, moving around is not in their repertoire, but the benefits of having a very fast signaling system when threatened by a predator are obvious nonetheless.

The Molecular Basis of the Shadow Reflex: The Light Response

The story of the molecular basis of vision begins in 1942 when Selig Hecht at Columbia University began experiments designed to ask whether human vision is sensitive enough to detect the smallest amount of light, a single quantum. (The answer is, yes, we can.) Aside from its interesting conclusion, one of the study's results was that Hecht's young graduate student, George Wald, became so fascinated with the question of how light is detected in the eye that he spent the rest of his scientific career identifying and studying the molecule that performs light absorption in the eye: rhodopsin. Rhodopsin is composed of the protein opsin linked to retinal, a small molecule derived from vitamin A. To everyone's surprise, rhodopsin was found to be ubiquitous in the animal world, enabling vision from jellyfish to humans. More surprising still was the subsequent discovery of a rhodopsin-like molecule in fungi and even in bacteria.

Like *Paramecium*, the cell membrane of the barnacle photoreceptor has specialized regions for particular functions (Fig. 2.9). The light-sensitive portion of the cell is at the end extending out from the brain and consists of extensive foldings of the

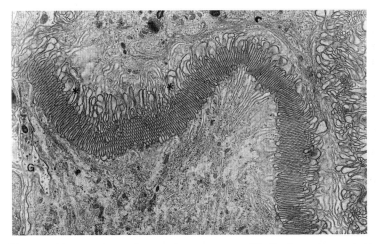

Figure 2.9. Photosensitive membrane from a photoreceptor cell in the horseshoe crab *Limulus*. The densely folded region of the plasma membrane contains rhodopsin and the proteins that perform the biochemical cascade following light absorption.

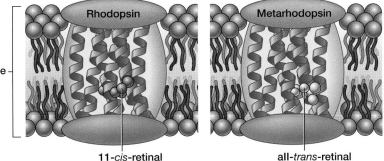

Figure 2.10. Conformational change in rhodopsin after light-induced isomerization of retinal. The change in shape of retinal induces a change of shape in the protein surrounding it, initiating the biochemical cascade of reactions.

membrane called microvilli. It is at these microvilli that the rhodopsin molecules are concentrated, embedded in the membrane.

A single molecule of rhodopsin is capable of absorbing a single quantum of light. When this occurs, it causes 11-*cis*-retinal to change its conformation to all-*trans*-retinal (Fig. 2.10). The new conformation does not fit well into the opsin protein and, as a result, causes the protein to change its conformation (Fig. 2.11). These shape changes are the starting point for a cascade of reactions consisting of conformational changes that activate enzymes to produce chemicals, which, in turn, produce conformational changes of other enzymes. These then produce chemicals, until ultimately, Na$^+$ channels open in the membrane, causing the light-stimulated photoreceptor to depolarize. Such is the case in all invertebrates. Vertebrates such as humans, however, have evolved a somewhat different light-response machinery (see Fig. 2.18 below). We use a similar rhodopsin, but the enzymatic cascade is different and our photoreceptors hyperpolarize, instead of depolarize, in response to light.

The opsin protein has a structure that is similar to a large family of membrane receptors, including many neurotransmitter receptors, all of which use a similar strategy for communicating with the inside of the cell. Each binds a small molecule within the plane of the membrane, causing a conformational change in the receptor so that it becomes able to bind onto a "G"-protein complex present nearby on the inner face of the membrane. The conformational change that occurs in this complex, involving the exchange of GDP for GTP (hence the name "G" protein), causes dissociation of the G-protein subunits, allowing one of the freed subunits to move along the membrane until it bumps into and activates a membrane-bound enzyme to initiate the internal signaling cascade (Fig. 2.12).

In the horseshoe crab *Limulus*, the steps in this cascade have been studied in some detail. There, the enzyme associated with the membrane is phospholipase-C

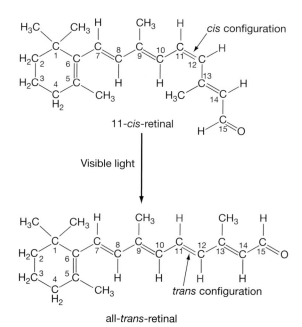

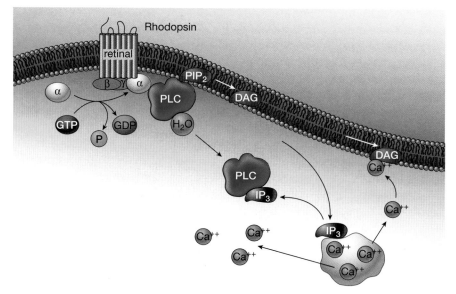

Figure 2.11. Light-induced isomerization of retinal. Shown in *blue* is the change from *cis* to *trans* configuration.

Figure 2.12. Intracellular signaling cascade triggered by rhodopsin–G-protein interaction (see text). (PLC) Phospholipase C; (PIP$_2$) phosphotidylinositol-4,5 bisphosphate; (DAG) diacylglycerol; (IP$_3$) inositol-1,4,5 triphosphate; (GTP) guanosine triphosphate; (GDP) guanosine diphosphate; (α, β, and γ) G-protein subunits.

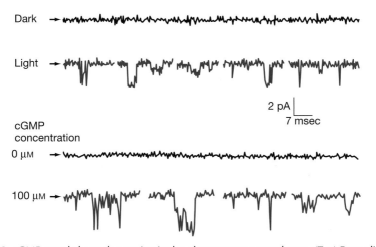

Dark

Light

2 pA

7 msec

cGMP
concentration

0 µM

100 µM

Figure 2.13. cGMP-gated channels opening in the photoreceptor membrane. (*Top*) Recordings of single channels opening in the photoreceptor membrane when light is on (*blue trace*) but not when it is off (*black trace*). (*Bottom*) Recordings of single channels opening in the presence of cGMP (*blue trace*) but not in its absence (*black trace*). These recordings were made in an isolated patch of the cell's membrane where single channel opening and closing can be detected. Channel opening produces a downward deflection, indicating that current (in pA) is flowing through the photoreceptor membrane.

(PLC). Thus begins the chemical messenger component of the cascade (Fig. 2.12) as PLC cleaves phosphotidylinositol-4,5 bisphosphate (PIP$_2$) into inositol-1,4,5 triphosphate (IP$_3$) and diacylglycerol (DAG). Both of these products act as internal messengers in the cell. A likely scenario of events in the *Limulus* eye is the following. IP$_3$ releases Ca^{2+} from internal compartments. Calcium then activates guanylate cyclase, which produces cyclic GMP (cGMP). cGMP binds directly to an Na$^+$ channel and causes it to open (Fig. 2.13), thus depolarizing the cell. A collateral signaling mechanism in these cells appears to involve DAG, the other product of PLC catalysis, activating another type of Na$^+$ channel. The two types of sodium channel are embedded in the membrane in the photosensitive part of the cell, the same region in which rhodopsin is found. The presence of two parallel systems for depolarizing the membrane recalls the degeneracy of mechanisms in *Paramecium* for repolarizing the membrane (Chapter 1). Such examples abound throughout nervous systems.

Signaling at the Synapse

Just as there are specializations for light sensitivity at one end of the photoreceptor cell, there are also specializations for the release of chemical signals at the synaptic end. These convert the electrical signal traveling down the axon into the trigger for the rapid secretion of neurotransmitter from the nerve terminal. The key player in this conversion is Ca^{2+}, which enters the terminal, when it is depolarized, through a specific class of voltage-sensitive calcium channels localized there. The sudden flooding of the nerve terminal with calcium activates the proteins involved in secretion of histamine. Histamine is contained in membrane-enclosed vesicles, and calcium catalyzes their fusion with the membrane of the nerve terminal. When the vesicles fuse with the outer membrane, their contents are dumped into the narrow space between the photoreceptor nerve terminal and the I cell. There, the neurotransmitter diffuses quickly and binds to its receptors on the I cell.

Like ion channels, much of the machinery for synaptic release is far more ancient than nervous systems themselves. The mechanism uses a system of membrane vesicle cycling that has many proteins in common with the apparatus involved in membrane trafficking and secretion in the cell's Golgi apparatus. Because many of these are present in unicellular protozoans, it seems likely that the machinery for synaptic release is derived from the eukaryotic cell's internal membrane system, which then became specialized for nerve terminals.

Synaptic vesicle fusion is a process that involves the binding and release of many proteins in order to shuttle a large mass of membrane to the outer plasma membrane. There, the lipid bilayers fuse and the neurotransmitter contents of the vesicle are expelled (Fig. 2.14). The steps of the sequence include "docking" of the vesicle on the inner face of the membrane and then a fusion event that depends on Ca^{2+}. The synaptic SNARE complex (soluble N-ethyl maleimide sensitive–factor attachment protein receptors) lies at the core of this process, consisting of syntaxin, synaptobrevin, and SNAP-25 proteins. First, SNARE complexes form and synaptotagmin

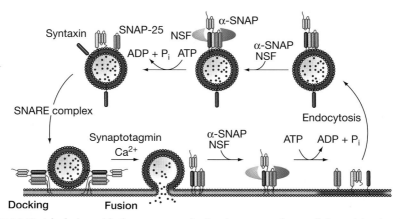

Figure 2.14. Vesicle fusing with the nerve terminal's plasma membrane, followed by the pinching off of a patch of membrane for recycling into a new vesicle. Shown are protein assemblies associated with vesicle release and recycling. The portion of membrane that is endocytosed is shaded *dark gray*. See text for description.

proteins associate with them. The assembly of this complex of proteins and the pressing of the vesicle against the membrane is thought to create an unstable intermediate in which the vesicle membrane is partially fused with the cell membrane. Ca^{2+} influx destabilizes this intermediate by triggering the Ca^{2+}-binding domains of synaptotagmins to associate with, and insert partially into, the lipid bilayer. The mechanical disturbance caused by this insertion then opens a pore in the incompletely fused membranes, which drives the process to completion. In the end, the vesicle membrane becomes fully integrated into the cell membrane (Fig. 2.14).

The sequence of events described above is probably not the only pathway of assembly that can lead to fusion. As described for ion channel mechanisms in *Paramecium* and in the barnacle phototransduction cascade, the simultaneous use of multiple, nonidentical (i.e., degenerate) mechanisms to accomplish a process is ubiquitous.

Without the foregoing mechanism of vesicle fusion, the barnacle photoreceptor would have no way of conveying its light-induced electrical signal to the postsynaptic I cell. The change in membrane potential would reach the terminal and fizzle out. In fact, the electrical signal does fizzle out at this point, but because it has triggered the release of the neurotransmitter histamine (Fig. 2.15) from its vesicles, the signal lives on. To instigate a membrane potential change in the adjacent cell, histamine binds specifically to receptors on the surface of the I cell that opens ion channels, allowing Cl^- to flow into the cell (Fig. 2.16). These are chloride-selective ion channels that hyperpolarize the cell when open.

A bit of the evolutionary history of histamine receptor/chloride channels can be inferred from their amino acid sequence. They are homologous to a wide range of chloride channels, some of which are not coupled to receptors, whereas others are coupled to receptors for neurotransmitters other than histamine. This suggests a scenario and history similar to the one described for potassium channels (Chapter 1), in which the ancestral protein is a pore and filter selective for that particular ion, coupled to a protein sequence that gives it a characteristic gating mechanism. In the barnacle's case, the filter is selective for chloride and the gating mechanism depends on the binding of histamine.

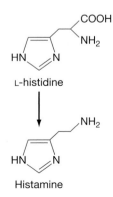

Figure 2.15. Histamine, neurotransmitter synthesized from histidine.

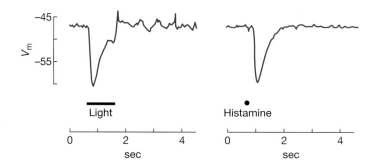

Figure 2.16. I-cell response to direct application of histamine. The application of histamine to an I cell's membrane (*right*) produces a change in membrane potential similar to that of a pulse of light (*left*).

Adapting to Different Light Intensities

During the day, when the barnacle is feeding, its photoreceptors are exposed to constant light. This means that the cell releases neurotransmitter constantly at a level proportional to the light intensity. In fact, it is crucial that the transmitter released from the photoreceptor be maintained at a relatively high level in the synapse so that any sudden reduction in transmitter can depolarize the postsynaptic cell and produce the alarm signal of the shadow response (Fig. 2.6). Because the cell has only a finite amount of histamine stored and a finite ability to replenish those stores when they are being constantly depleted, the cell needs a mechanism to reclaim the expelled transmitter. This exists in the form of a histamine transporter protein—a membrane protein that selectively pumps molecules of histamine from the synaptic space back into the cell. Such transmitter-specific transporters exist for virtually all neurotransmitters.

By the same token, the synapse must maintain its ability to respond sensitively to any change in illumination regardless of the starting level. In other words, the barnacle must recognize a shadow whether it occurs in dim light or in bright light. As an apparent adaptation to this requirement, the level of transmission at the synapse adapts to the existing, stable level of light intensity by tempering the histamine response if it persists long enough (Fig. 2.17). By increasing its rate of histamine reclamation, the histamine transporter maintains the ability of the terminal to release transmitter over a wide range.

If it is beginning to appear that regulation can occur at any site in the cell or at any stage of a signaling cascade, or for any protein, this is no accident. One of the hallmarks of evolution is the acquisition of mechanisms to adjust, tweak, and nudge the cell's many processes at virtually any possible time or place. Not all such sites will actually be capable of regulation in any one cell or species, but sooner or later, if we search long enough, virtually any case of regulation that we imagine can be found.

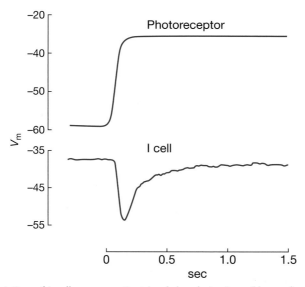

Figure 2.17. Adaptation of I-cell response. Sustained depolarization of barnacle photoreceptor cell (*upper trace*) produces a strong initial response in the postsynaptic I cell, which then gradually falls back to its baseline value (*lower trace*).

The Evolution of Eyes

The question of the origins of the animal eye has a long history of challenging and fascinating biologists, including Darwin himself. In recent times, molecular comparisons have added to the intrigue by showing that even in organisms that diverged as long ago as fruit flies and mice, remarkably similar genes (e.g., *Pax6*) are involved in establishing cell fate in the early developmental events that produce an eye. A similar degree of conservation is found in the rhodopsins that initiate their light responses.

But fly eyes, which are very similar to *Limulus* and mouse eyes, which are also very similar to our own, differ from one another in many significant ways. Not the least of their differences are in the signaling cascade and ion channels that mediate the photoresponse after rhodopsin absorbs light (Fig. 2.18). The two types of eye are also very different in structure. The vertebrate eye has a single large lens that forms an image onto the entire sheet of photoreceptor cells, all of which are organized into a single layer. The arthropod compound eye (e.g., insects, horseshoe crabs, crustaceans, spiders) is a structure comprised of many repeating cartridge-like units (ommatidia), each of which has its own little lens and set of several photoreceptor cells. Both types of eye also have pigmented cells associated with the photoreceptors, a parallel layer in the vertebrate eye, and a lining of cells within each ommatidium of the arthropod eye (Fig. 2.18).

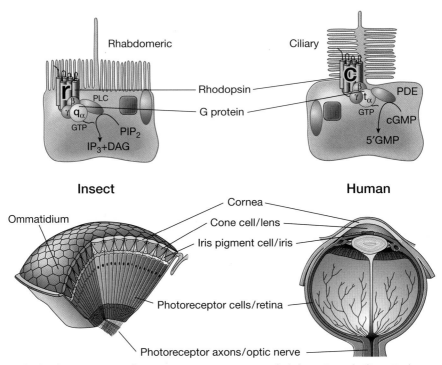

Figure 2.18. Photoreceptor cells come in two major types: rhabdomeric and ciliary. Each type has characteristically different opsins, G proteins, and signal transduction systems using different chemical messengers: phosphotidylinositol-4,5 bisphosphate (PIP_2) and its associated enzyme phospholipase C (PLC) in rhabdomeric photoreceptors (*top left*), and cGMP with its associated phosphodiesterase (PDE) in ciliary photoreceptors (*top right*). Insect compound eyes (*bottom left*) are composed of rhabdomeric photoreceptor cells. Vertebrate eyes consist predominantly of ciliary photoreceptors (*bottom right*). Vertebrate eyes generally have camera-like lenses forming a single image onto a sheet of photoreceptors, whereas compound eyes have multiple compartments (ommatidia), each with its own lens and small set of photoreceptors. For other abbreviations, see Fig. 2.12.

Vertebrate and arthropod eyes, however, are far more elaborated and intricate than those of many animals from lower branches of the phylogenetic tree (known as lower Metazoans), in which photosensitive cells are nonetheless widespread. Cnidarians, the phylum that includes jellyfish, hydra, and coral, are the simplest creatures that contain all of the machinery of a nervous system (see Chapter 3). Their light-sensitive organs run the gamut from single photosensitive cells to full-blown eyes with single, image-forming lenses (Fig. 2.19). Image-forming lenses are rare among invertebrates, appearing only in jumping spiders, octopi, and a few jellyfish. The first two of these species are well known for the visual acumen that they use in hunting. Jellyfish are not. But box jellyfish use their "eyes" for navigating their way among underwater obstacles, such as the pylons of a pier or the tangled roots of a mangrove swamp. Remarkably, a

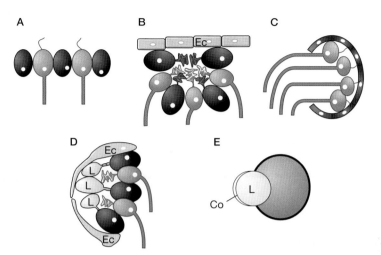

Figure 2.19. Cnidarians have nearly the full range of eye types in one phylum, from primitive eye spot to camera lens eye. Photoreceptor cells are shown in *blue*; pigment cells are *black*. (Ec) Epithelial cells; (L) lens; (Co) cornea. (A) Primitive eye spot in the hydrozoan *Leuckartiara*. (B) Everted optic cup closed by a layer of epidermal cells in hydrozoans *Polyorchis* and *Bougainvillia*. (C) Inverted pigment cup of scyphozomedusae *Aurelia*. (D) *Cladonema* ocellus with lens bodies formed as cytoplasmatic extensions of individual pigment cells. Epithelial cells extend over the lens bodies to form a primitive "cornea." (E) Complex camera-type eye of the cubomedusan *Tripedalia*, containing cornea, lens, retina, and pigment layer.

homolog of the *Pax6* gene is present in these creatures and marks the precursor cells that will give rise to the sensory structure containing its "eye." Genes of the *Pax* family (including *Pax6* itself) have been shown to be present in precursors of other sensory cells in addition to photoreceptors, and this is true of the jellyfish *PaxB* gene as well. The sensory structure containing the "eye," called the rhopalium (Fig 2.20), also contains mechanosensory cells and additional, simple photoreceptor organs consisting of one or a few cells. The *PaxB* gene is expressed in the rhopalium sensory cell precursors.

Cnidarians, however, are not the simplest animals that have photosensitive behavior. Even the sedentary, neuronless, and presumably blind sponges (Porifera) have *Pax* gene homologs, and some sponge larvae respond to light with altered swimming patterns. These *Reniera* larvae have light-sensitive cells at the base of their cilia. The photopigment is not rhodopsin, but flavin, a molecule derived from B vitamins that is involved in various cellular reactions. The flavin-containing protein is unrelated to the opsins, but instead is closer to a class of proteins that includes cryptochrome, a protein involved in circadian rhythms (see Chapter 5). The mechanisms of the photoresponse in these organisms, or in any of the Cnidarians for that matter, remain unexplored.

Nor are Cnidarians and Poriferans the simplest animals in which *Pax*-like genes play a role in the development of their neural or neuroid structures. Placazoans, represented by the lone species *Trichoplax adhaerens*, are the simplest of all multicellu-

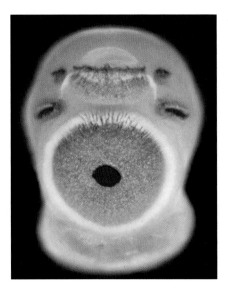

Figure 2.20. Rhopalium, the complex sensory structure in cubomedusan (box jellyfish) *Tripedalia*. The camera lens eye is the large structure in the center, the structures above it are simpler photoreceptors, and the portion of the rhopalium below the large eye is a mechanosensory organ.

lar animals, and are known as Mesozoans. They lack neurons and have only four different cell types, none of which are readily recognizable as being homologous to Metazoan cell types. The *Trichoplax* version of a *Pax* gene, *PaxT*, is present in the precursor cells to one of these layers. This cell layer is thought to mediate coordinated shape changes that *Trichoplax* executes when it is ingesting food. Thus, the *Pax* gene in the simplest multicellular animal looks as though it is associated with cells that do for *Trichoplax* what neurons and muscles do for Metazoans. Whether *Trichoplax* represents an example of what the precursors of nervous systems looked like remains one of those intriguing but ultimately unanswerable evolutionary questions.

Further Reading

Arendt D. 2003. The evolution of eyes and photoreceptor cell types. *Int. J. Dev. Biol.* **47:** 563–571.

Dorlochter M. and Stieve H. 1997. The *Limulus* ventral photoreceptor: Light response and the role of calcium in a classic preparation. *Prog. Neurobiol.* **53:** 451–515.

Gwilliam G.F. and Millecchia R.J. 1975. Barnacle photoreceptors: Their physiology and role in the control of behavior. *Prog. Neurobiol.* **4:** 211–239.

Stuart A.E. 1999. From fruit flies to barnacles, histamine is the neurotransmitter of arthropod photoreceptors. *Neuron* **22:** 431–433.

Stuart A.E. and Oertel D. 1978. Neuronal properties underlying processing of visual information in the barnacle. *Nature* **275:** 287–290.

CHAPTER 3

Truth for the Jellyfish
Coordination to Fit the Occasion

> Truth on our level is a different thing from truth for the jellyfish, and there must certainly be analogies for truth and error in jellyfish life.
>
> T.S. Eliot

The first fossil evidence for animal life on earth appears in geological strata dating back 600 million years in the Precambrian Proterozoic era. They are recognizable as animals because some of them resemble jellyfish (Fig. 3.1), although not all paleontologists agree that they are. These jellyfish-like creatures are the predominant representatives of multicellularity in that epoch (all 80 million years worth), or at least the major ones to have been preserved. Fossils from the Cambrian era look even more like jellyfish. Unfortunately, it is impossible from these fossilized impressions to

Figure 3.1. Jellyfish-like fossils from 600 and 540 million years ago. (*Left*) *Mawsonites spriggi* found in Australia and (*right*) *Maotianoascus octonarius* from China.

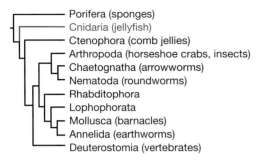

Figure 3.2. Phylogenetic tree depicting relatedness among animals based on gene sequences. Cnidaria is the earliest branching group that has a nervous system.

determine what (if any) kind of nervous system they had. It is equally impossible to know what their predecessors had in the way of nervous systems. DNA sequence analysis agrees with the evidence from the fossil record on the primitiveness of jellyfish in placing them and other Cnidarians near the base of the Metazoan (animal) tree (Fig. 3.2). In favor of this placement is the presence in their genomes of the full range of metazoan genes for signaling pathways as well as a substantial number of genes otherwise found only in plants and bacteria (Fig. 3.3).

Although it is unsafe to assume that existing jellyfish are like their Precambrian or Cambrian forebears—the tribulations of surviving for 600 million years could have brought about many changes in the Cnidarian genome—it is reasonable to assume that they give us as good a picture as we will ever get of how the first nervous systems and the behaviors they produced might have looked. One of the notable features of the jellyfish nervous system is that, despite its primitiveness, it is fully equipped with all of the essential machinery (e.g., ion channels, neurotransmitters, synapses, receptors, transporters). Hence, jellyfish allow us to ask what kinds of behavior and neural function can be generated by this machinery in a relatively simple form.

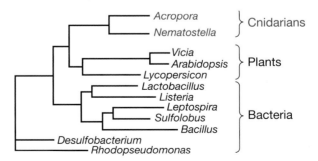

Figure 3.3. Phylogenetic tree depicting sequence similarity for a bacterial gene (called *UspA*) among Cnidarians, protozoans, plants, and bacteria. *UspA* is one of many genes common to Cnidarians, protozoans, plants, and bacteria that are absent in other animals.

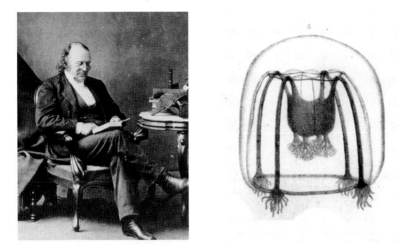

Figure 3.4. Louis Agassiz (1807–1873) and the jellyfish *Bougainvillea* that he studied.

Historically, based on the lack of a centralized brain, jellyfish were considered not to have any nervous system at all. Louis Agassiz, the Swiss naturalist who founded Harvard's Museum of Comparative Zoology, identified the first Cnidarian nervous structures in *Bougainvillea superciliaris* in 1850 (Fig. 3.4), although the concept was not accepted until German zoologist Ernst Haeckel's studies several decades later.

Alternative Ways of Swimming

The jellyfish *Aglantha digitale* live their lives in the cold waters of northern oceans, drifting or swimming to change direction or depth. When they encounter food, in the form of tiny marine crustaceans (copepods) swimming along, their tentacles carry it to their centrally located mouth (which is the same orifice used for excretion in these radially symmetrical creatures).

They swim by forcing water out of their bells. Water expulsion is produced by contraction of a conical sheet of muscles, called the myoepithelium, inside the umbrella of the bell. These muscle contractions are driven, in turn, by eight motor neurons that are part of the parallel rings of neurons that encircle the base of the bell (Fig. 3.5). The amplitude, frequency, and directionality of this maneuver determines what kind of swimming they display, from relaxed "slow" swimming with its "fishing" behavior, to their rapid escape response. Despite the simplicity of their nervous systems (~1000 neurons), they display a remarkable degree of agility and flexibility in their behavior.

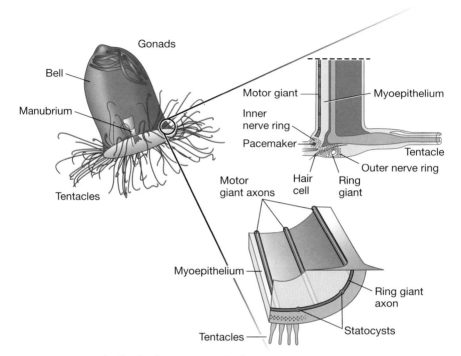

Figure 3.5. Anatomy of *Aglantha* showing major body structures (*left*) and arrangement of the major bundles of neurons (*right* and *below*, in *gray*).

Slow Swimming

In slow swimming, *Aglantha* contracts its bell rhythmically for several cycles, then floats passively downward with its tentacles spread out like a net, "fishing" for floating tidbits. When it has floated down for a minute or two, it rights itself and resumes the rhythmic swimming cycle upward for a minute or so (Fig. 3.6). Rhythmic swimming is under the control of a pacemaker system of neurons that form one of the nerve rings around the base of the bell (Fig. 3.7). The pacemaker neurons fire action potentials rhythmically at intervals of approximately 2 seconds during the swim cycle that result in a contraction of the bell once every 2 seconds (Fig. 3.8). These pacemaker cells keep time for the jellyfish, and they communicate this rhythm through synapses onto a series of "giant" motor neurons (so called because of their relatively large [30–40-µm] diameter) that extend from the base of the bell up into the myoepithelium (Fig. 3.7). The motor neurons produce action potentials that activate the sheet muscles to contract and expel water from the bell. These mild contractions propel the animal forward (or upward, as the case may be) by about one body length.

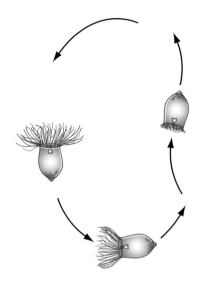

Figure 3.6. "Fishing" behavior in *Aglantha*.

At the end of the downward drift in their "fishing" behavior, *Aglantha* rights itself (Fig. 3.6). This response is stimulated by a set of eight gravity-sensing, bulb-shaped structures (statocysts) lining the rim of the bell (Fig. 3.5) that bend when the animal tilts. When its statocysts are removed, *Aglantha* swims around aimlessly with no discernible orientation. These structures are covered with mechanosensory neurons that generate changes in membrane potential when touched by the bending of the statocyst, which deforms the membrane and opens mechanosensory channels. (The

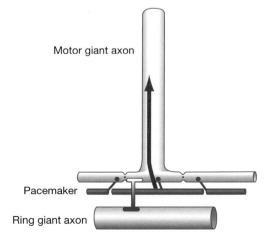

Motor giant axon

Pacemaker

Ring giant axon

Figure 3.7. Pathway of signaling (*gray arrow*) from the neurons comprising the pacemaker system to motor giants involved in slow swimming behavior in *Aglantha*.

Motor giant

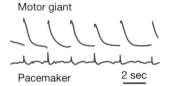

Pacemaker   2 sec

Figure 3.8. Rhythmic activity of the pacemaker system (*bottom*) driving motor giant neurons (*top*).

mechanosensory channels on the statocysts are functionally similar to those in *Paramecium*, described in Chapter 1. Statocysts also share many features with the hair cells in our middle ear that give us our sense of balance.)

The mechanosensory neurons in the statocysts synapse onto the largest neurons in the animal, the "ring giants" that form one of the major concentric nerve rings around the base of the bell (Fig. 3.6). The ring giants, in turn, synapse onto the giant motor neurons innervating the myoepithelium. It seems likely that when the statocysts are differentially stimulated, as during a tilt, their deformation stimulates the myoepithelium asymmetrically, resulting in a righting response. In other words, each portion of the bell responds independently to its local statocyst.

Escape Swimming

When *Aglantha* brushes against a predator, or receives a bite from a mackerel, it executes an escape response. This consists of several, rapid (80 msec), powerful contractions of the bell that propels it forward much faster and further than during slow rhythmic swimming—up to five body lengths per contraction at a speed of 40 cm/sec. During these contractions, nearly all of the water is expelled from the bell, 80% of it within the first 20–30 msec. This response overrides anything else the animal is doing.

The sensory cells that trigger this response are located on the tentacles, which are evenly distributed around the margin of the bell (Fig. 3.5). Near the base of each tentacle is a cluster of hair cells (Fig. 3.9), similar to those on the statocyst. A strong stimulus to the tentacle causes the hair cells to initiate a burst of electrical activity that is communicated to the ring giant axon through synapses (Fig. 3.10). The ring giant is a continuous cable running around the bell margin, consisting of one long, multinucleated axon. Once in the ring, the signal is propagated as an action potential that circles around in both directions from the point of the stimulus. (This raises two interesting metaphysical questions: First, what happens when two action potentials going around the ring in opposite directions collide with each other? Answer: They cancel each other out. Second, if an action potential started off in only one direction, what would prevent it from going around the ring indefinitely? Answer: The mechanisms for repolarizing the membrane, which terminate an action potential [see Chapter 1], prevent this from happening.)

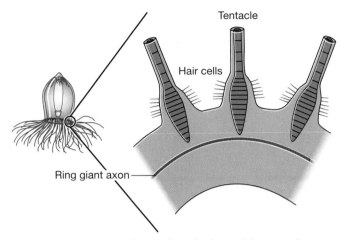

Figure 3.9. Sensory hair cells at the base of the tentacles.

As the action potential circumnavigates the bell, it drives the ring giant's synapses to release neurotransmitter, and this stimulates the motor giant neurons in the myoepithelium (Fig. 3.11). *Aglantha* uses this three-step, rapid communication system to ensure that the alarm signal is sent rapidly, reliably, and widely around the bell, through the ring giant, through the motor giants, to the muscles that contract the bell (Fig. 3.11). The action potentials are conducted rapidly enough to produce near synchrony in the activation of the bell muscles. As an extra measure of coordination, the cells of the sheet muscles in the myoepithelium are also directly coupled to one another. This coupling is achieved by means of junctions in the membranes where myoepithelial cells contact one another. The junctions allow direct electrical communication among the cells that, as a result, permits action potentials to spread from one muscle sheet to another. This further ensures a nearly simultaneous response all around the bell.

Mixed Signals from Mixed Ion Channels

You may have noticed that both swimming responses, slow and escape, use motor giant neurons propagating action potentials to activate the myoepithelium. How can the same neurons, using action potentials, produce mild muscle contractions during slow swimming and massive muscle contractions during escape swimming? This is especially puzzling, given that action potentials are stereotypical—rising to the same peak time after time (see Chapter 1).

The answer is found in an unusual adaptation that the motor giant neurons of *Aglantha* have evolved that enables them to conduct two different sizes of action potential (Fig. 3.12). Which size they produce depends on the strength of the signal they receive, and the signal strength differs between the two sources of signal to the

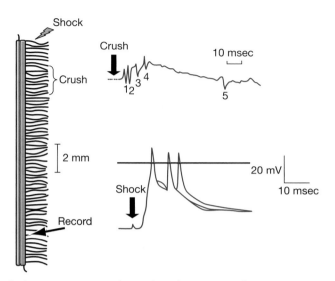

Figure 3.10. Multiple receptor potentials (numbered on *upper right trace*) are generated in the ring giant axon by crushing hair cells (*"Crush" arrow*) on the bell margin (*left*) or by delivering an electric shock (*"Shock" arrow*) to the hairs. Action potentials are sent around the ring giant axon after an electric shock (*lower right trace*).

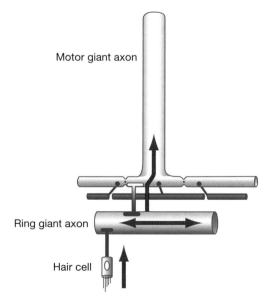

Figure 3.11. Pathway of signaling (*gray arrows*) for triggering an escape response from hair cell to ring giant to motor giant axon.

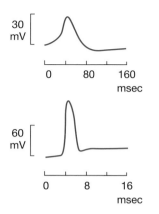

30
mV

0   80   160
         msec

60
mV

0   8   16
        msec

Figure 3.12. Smaller Ca-dependent action potential (*top*) and larger Na-dependent action potential (*bottom*) in motor giant axon.

motor giants: the pacemaker system for slow swimming and the ring giant for escape behavior. The potentials that result from pacemaker neurons are relatively slow to rise and relatively small at their peak. They depolarize the motor giant membrane potential from its resting value of −70 mV up to only −51 mV. This amount of depolarization activates a class of voltage-sensitive calcium channels in the motor giants to trigger an action potential. The peak of these calcium action potentials is also relatively small (30 mV from base to peak; see Fig. 3.12) and when they reach the motor terminals, they stimulate a limited amount of transmission at their synapses and thus a limited amount of depolarization in the myoepthelial cells.

In contrast, the synaptic potentials that are evoked in the motor giants by input from the ring giant axon, which are set off by a strong stimulus such as disturbing a tentacle, are faster and larger. They depolarize the motor giant membrane to a higher level (−32 mV), which activates another class of channels to trigger a faster and larger (100 mV amplitude) action potential in the motor giants (Fig. 3.12). These more potent action potentials use sodium channels triggered by the greater depolarization. The effect of this stronger stimulus on synaptic transmission to the myoepithelium is correspondingly faster and larger, producing stronger contractions in the muscles that spread throughout the muscle sheet. This spread is helped along by the membrane junctions between myoepthelial cells that allow the free passage of electric current. (The weaker depolarizations in slow swimming also spread, but because they are weaker, they spread less and produce less contraction.)

But a question still remains as to why the calcium action potentials are so much smaller than those based on sodium ions. Why don't the calcium action potentials produce enough of a depolarization to trigger the sodium channels? Surely the concentration difference for calcium between the inside of the cell and the outside is not so miniscule that it fails to produce a major voltage difference.

Answering this question led to the discovery of yet another special adaptation in these motor neurons: a mechanism for selectively short-circuiting the slower, small-

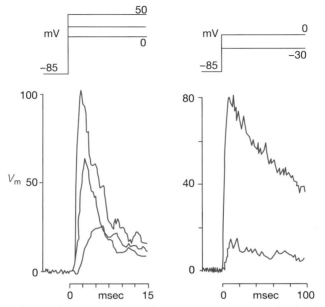

Figure 3.13. Two classes of potassium channels in motor giant axons of *Aglantha*, one of which opens and closes faster (*left*) than the other (*right*). Traces (*blue*) show superimposed recordings from single K⁺ channels generated by graded depolarizations (*top black traces*) of the membrane. Stronger depolarizations evoke a larger response.

er calcium action potential. *Aglantha* accomplishes this by means of a special class of potassium channels whose response properties match those of the calcium action potential. These channels open when the membrane potential has not depolarized very much, and their slow opening matches the rise of the calcium spike. This potassium channel could also, in principle, short-circuit any action potential, but it opens too slowly to catch up with the much more rapid sodium action potential (Figs. 3.12 and 3.13). The effect of these channels is to cut off the calcium action potential before it has reached what would be its peak based on the external calcium concentration.

Because action potentials all use potassium channels to repolarize the membrane (see Chapter 1), a different set of them must be participating in the faster sodium action potentials associated with the escape response. This was confirmed by identifying another class of these channels that do not start to open until the membrane has reached a much higher level of depolarization, which occurs during a sodium spike. They are also fast enough to keep up with the quicker rise of the sodium spikes.

Aglantha's strategy for generating two different responses in the same cells thus depends first on the motor giants receiving differently sized synaptic signals, and second on keeping the peak of one type of action potential (calcium from a weak stimulus) too low to trigger the other type of action potential (sodium from a stronger stimulus). This keeps the two kinds of action potentials separate and makes it possible for them to be used to communicate signals of different strength and time course.

Ion Channels in Jellyfish

Having already seen that you do not need a nervous system to have ion channels (Chapter 1), there is no reason to expect that the ions channels responsible for *Aglantha*'s swimming should be unusual. Actually, we do not yet know what *Aglantha*'s channels look like, because none of its genes have been cloned. In other jellyfish, however, we do know that their voltage-sensitive calcium and sodium channels bear a family resemblance to those in other animals. When the sequences of a wide sampling of these channels from many organisms are carefully examined and compared, it appears in fact that there are vague resemblances among all of the channel types (Fig. 3.14). Potassium channels are likely to have been the first, being structurally the simplest. Calcium channels appear to have evolved from them by a process in which separate potassium channel subunits became fused into progressively longer proteins that now had multiple transmembrane domains. Sodium channels probably evolved as variations on these calcium channels.

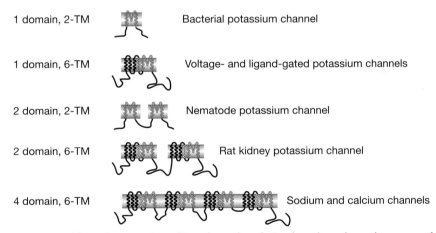

1 domain, 2-TM		Bacterial potassium channel
1 domain, 6-TM		Voltage- and ligand-gated potassium channels
2 domain, 2-TM		Nematode potassium channel
2 domain, 6-TM		Rat kidney potassium channel
4 domain, 6-TM		Sodium and calcium channels

Figure 3.14. Hypothetical progression of ion channel evolution based on channel structures found in existing species. The simplest potassium channel protein is found in bacteria, with a single domain having two transmembrane (TM) segments. Most voltage- and ligand-gated potassium channels are single domain with six transmembrane segments. A two-domain potassium channel of the 2-TM type is found in the nematode *Caenorhabditis elegans*, and a two-domain 6-TM type is found in rat kidney. Sodium and calcium channels have four domains, each with six transmembrane segments. (*Gray*) Primordial channel domain, (*black*) voltage-sensing domain (see Chapter 1); (*blue*) membrane.

Neuronal Timekeeping

The foregoing discussions of swimming rhythm, escape behavior, and channel properties raise the all-important issue of timing in the nervous system. Whereas timing is
obviously important in a jellyfish nervous system, careful control over timing took on
enormous importance when nervous systems evolved up to the size of several hun-
dred thousand neurons (as in flies) or tens of millions (as in rodents).

Timekeeping in the nervous system can be the property of a single neuron,
where it depends mainly on the kinetics of channel opening and closing, or it can be
a network property of neuronal circuitry, where it can depend on a variety of different effects, separately or in combination.

Pacemaker Neurons

The timekeeper for *Aglantha*'s slow, rhythmic swimming is its system of pacemaker
cells with their regular train of depolarizations (Fig. 3.6). Neurons of this type are
widespread in the animal world and have been found in mollusks and insects, as well
as in our own eyes, hearts, and brains. In the sea snail *Aplysia*, the properties of these
cells have been studied in some detail.

Alplysia californica lives in the intertidal zone along the California coast where
they crawl around, eat seaweed, and mate (generally in groups). Their nervous system consists of a handful of ganglia (clusters of neurons), one of which, the abdominal ganglion, regulates their digestion and water regulation. A cell known as R15 in
this ganglion is a pacemaker cell that displays "bursting" activity (Fig. 3.15). This
means that it fires bursts of action potentials at regular intervals, whenever its membrane potential is in the right range. Below −70 mV the cell is silent and above −30
mV the cell fires action potentials continuously. Underlying this bursting behavior is
a slow oscillation of the resting membrane potential that periodically takes it above
the threshold for action potentials and then brings it back down below that threshold. This oscillation is the key to the cell's pacemaker activity.

An increase and decrease in the Ca^{2+} concentration inside the cell accompanies
the oscillation in membrane potential (Fig. 3.16). This is due to a type of calcium
channel that is regulated by both voltage and calcium. The channels' voltage sensitivity induces them to open as soon as depolarization begins, and their calcium sensitivity causes them to close after enough calcium has entered the cell. (They are partially

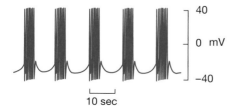

40

0 mV

−40

10 sec

Figure 3.15. Bursting activity of the R15 neuron
of *Aplysia*, consisting of trains of action potentials
with quiet, ~10-sec intervals between clusters.

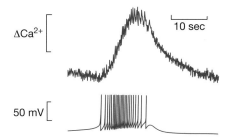

ΔCa^{2+}

10 sec

50 mV

Figure 3.16. A slow oscillation in the concentration of intracellular calcium underlies bursting behavior in the R15 neuron. (*Top trace*) The rise and fall of intracellular Ca^{2+}, directly measured by a calcium-sensitive electrode; (*bottom trace*) concurrent firing of action potentials in the cell.

ligand-gated, as described in Chapter 1, where binding calcium causes them to close up.) Reinforcing this cycle, the rise in intracellular calcium also opens a class of calcium-sensitive potassium channels. As calcium builds up in the cell during bursts, the calcium-activated potassium channels are increasingly stimulated to open, finally hyperpolarizing the membrane enough to terminate the set of action potentials.

The bursting cycle occurs more slowly than the action potentials (Fig. 3.16). The process restarts after transporters pump out the excess calcium from inside the cell, which lowers its concentration enough to remove the source of inactivation for the calcium channels so that they can begin to open again. As they open and start to depolarize the cell, their voltage sensitivity causes them to open further, more calcium enters, and the cycle repeats itself.

Differential rates of channel opening and the combining of opposite tendencies are key to R15's pacemaker activity. In addition, the duration of bursts, the intervals between them, and the spike frequency in a burst are all functions of the mix of channels and the tuning of those channels. Depending on how these factors are combined, a wide range of firing characteristics is possible.

Circuitry and Timing

The swimming strategy of *Aglantha* grows out of its body plan. The radial symmetry of its bell allows its musculature to contract as a unit. Most animals, on the other hand, are bilaterally symmetrical, and thus have evolved different strategies matched to their anatomy. Swimming in the medicinal leech *Hirudo medicinalis* involves a carefully choreographed pattern of muscle contractions along its body wall. The result is a wave of movement that depends on its segmented anatomy, elongated shape, and bilateral symmetry (Fig. 3.17). The sequencing of neuronal activities that leeches use for this behavior extends much more widely than the channels in a single cell. The strategy is found in the nervous system's circuitry and its coordinated use of timing.

Leeches swim in their pond and river habitats in pursuit of a meal or to elude potential predators. They perform body bends while swimming that result from the opposing action of muscles on the dorsal (top) and ventral (bottom) body wall. When the ventral muscles in one segment contract and the dorsal muscles immediately above them relax, that portion of the body will bend downward in a U shape (Fig.

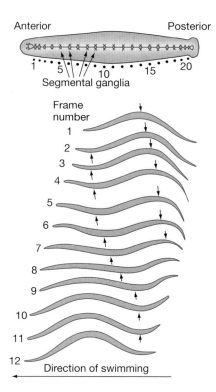

Figure 3.17. (*Top*) Segmented anatomy of the leech showing nerve cord (*blue*). Bend propagation along leech body during swimming (*bottom*). *Arrows* indicate positions of bends in sequential frames.

3.17). As the combination of ventral contraction and dorsal relaxation moves sequentially from segment to segment down the length of the body, the bend travels and produces a wavelike movement. Each segment has its own set of muscles and its own ganglion that are generally similar from segment to segment (Fig. 3.18).

The key to the opposing contractions of the dorsal and ventral muscles in each segment is in the antagonistic action of the leech's motor neurons. Each dorsal segment has excitatory and inhibitory motor neurons, as does each ventral segment (Fig. 3.19). One mechanism for ensuring coordination can be traced to the fact that the inhibitory neurons not only synapse onto the muscles, but also onto the excitatory neurons. That means that when the dorsal muscles are inhibited, so are the dorsal exciter neurons, keeping everything consistent. Similarly, on the ventral side, inhibition of the ventral muscles is accompanied by inhibition of the ventral exciter neurons. (An example of inhibitory synaptic transmission, the hyperpolarization of the barnacle's I cell by its photoreceptor, is described in Chapter 2.)

As the wave of contraction moves from segment to segment down the leech's body, the activity of any motor neuron in a given segment is "on" some of the time and "off" the rest of the time. Like the R15 cell in *Aplysia*, the membrane potential of excitatory motor neurons rises and falls with their bursts of activity during the swimming rhythm (Fig. 3.20). Unlike R15, however, their oscillating activity is not

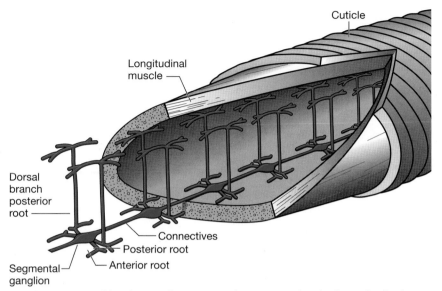

Figure 3.18. Anatomy of leech central nervous in the context of its body (with all other internal organs not shown). Segmental ganglia are shown along the ventral (lower) surface of the body cavity, sending branches to the ventral surface and up to the dorsal (upper) surface. These branches include motor neurons that synapse onto the longitudinal muscles of the body wall.

due to an internal pacemaker mechanism, but to the circuitry—the network of neurons synapsing onto each other—that drives the inhibitory motor neurons on the opposite side to be active or inhibited. But this begs the question of how the firing of the inhibitory neurons becomes periodic. The answer to this question lies in the tangle of neurons found in the segmental ganglion in the middle of each segment.

A subset of these neurons, including those that form synapses onto the dorsal and ventral inhibitory motor neurons, constitute the "oscillator" and set the timing of contractions in that segment (Fig. 3.21). These neurons fire in time with the swimming

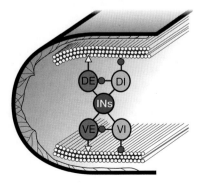

Figure 3.19. Simplified diagram of excitatory and inhibitory motor neurons on one side of a representative segmental ganglion of the leech. Muscles on the dorsal and ventral surface are shown being contacted by a representative excitatory (DE, VE) and inhibitory (DI, VI) motor neuron. (*Blue circles*) Inhibitory synapses; (*white triangles*) excitatory synapses. The complex of ~400 neurons in the segmental ganglion (INs) connects to the dorsal and ventral motor neurons. When dorsal muscles are inhibited by DI, their excitatory motor neurons are also inhibited. VI plays a similar role on the ventral side.

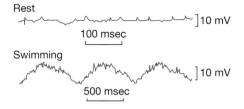

Rest

100 msec

]10 mV

Swimming

500 msec

]10 mV

Figure 3.20. Rhythmic activity of an excitatory motor neuron in a leech segment as it drives a muscle required for bending the body in swimming. The neuron shows a low level of activity at rest (*upper trace*) and an oscillating membrane potential that results in bursting activity during swimming (*lower trace*).

rhythm. This in itself would not necessarily assure them a role in setting the pace; they may just be following it. Their active role is shown by their ability to shift the timing of the swimming rhythm if stimulated directly, that is, if depolarized or hyperpolarized by the experimenter (Fig. 3.22). In other words, depolarizing one of these neurons for 2 seconds delays the motor neuron firing sequence by the same interval. This finding not only establishes their importance to rhythm generation, but also shows that the rhythm does not originate from any individual cell, but must be a product of the interconnected network. This is the secret to how nervous systems produce complex patterns of activity.

Inside one of the segmental ganglia, nearly all of the neurons' synapses are inhibitory, that is, neurotransmission at their synapses inhibits the activity of the neighboring, postsynaptic cells. Many of these neurons, in fact, inhibit one another reciprocally (Fig. 3.23). This may seem counterproductive, as if these cells are all trapped in repressive relationships, but in fact, they have more latitude owing to the arrangement of all of the connections in the network (Fig. 3.23). Although the exact

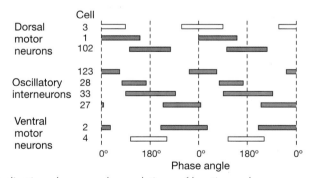

Figure 3.21. Coordination, shown as phase relations, of bursting cycles among motor neurons and interneurons during swimming within a single leech segment. Representative cells of three different types (dorsal motor neurons, interneurons, and ventral motor neurons) are identified by numbers to the right of the cells. (*Gray*) Interneurons; (*blue*) inhibitory motor neurons; (*white*) excitatory motor neurons. Each horizontal bar represents the duration of that cell's bursting cycle. A full cycle (0°–360°) can last 0.4–2 sec. Rhythmic sequences of advancing activities ensure the orderly contraction and relaxation of segmental muscles on the dorsal and ventral surfaces. Excitation of muscles on one surface is coincident with inhibition of muscles on the opposite surface.

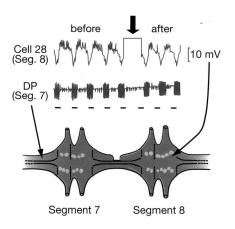

before   after

Cell 28
(Seg. 8)

[10 mV

DP
(Seg. 7)

Segment 7   Segment 8

Figure 3.22. Neuronal timing sets the swimming rhythm. (*Upper trace*) Rhythmic bursts of action potentials in interneuron 28 from segment 8 (*right arrow*). (*Lower trace*) Bursts of action potentials from segment 7, recorded from the main nerve (DP) exiting the ganglion (*left arrow*). When interneuron 28 is artificially depolarized (*square wave; upper arrow*), its firing phase is reset, and with it the firing phase of the DP nerve in segment 7. Timekeeping by interneuron 28 is a function of its place in the circuitry of the segmental ganglion.

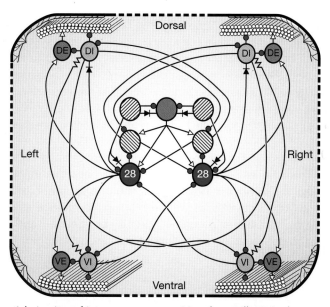

Figure 3.23. Essential circuitry of interneurons comprising the oscillator in the segmental ganglion. (DE) Dorsal excitatory motor neuron; (DI) dorsal inhibitory motor neuron; (VE) ventral excitatory motor neuron; (VI) ventral inhibitory motor neuron; (28) interneuron. (*Blue circles*) Inhibitory synapses; (*white triangles*) excitatory synapses; (*black triangles/zigzags*) electrical coupling. Cells are arranged to highlight their network connections, not according to actual anatomy. This illustration represents a small fraction of the ~400 neurons in each segmental ganglion.

nature of the interactions responsible for the oscillations remains to be defined, the alternating activity of these motor neurons produces a cycle of inhibition and release in the excitatory motor neurons of each segment. When they are in the released phase, they fire actively and cause muscle contractions.

The final link in the swimming rhythm is movement of the antagonistic contractions down the leech's body from segment to segment in a smooth progression. The advancing of the contraction cycle occurs when coordinating signals are sent from the oscillator neurons of one segment to oscillator neurons in the adjacent segment. These signals from segment to segment are conveyed by action potentials carried by bundles of neurons (called "connectives") that, when they arrive, shift the phase of the oscillator in the adjacent segment. In this way, each segment coordinates the cycle in the next segment ahead of it, and contractions are advanced along the body.

As in all biological explanations, there is much more going on here than is represented in the foregoing images and descriptions—more cells, more connections, more complex interactions. But these examples of swimming in *Aglantha* and the leech serve to illustrate the possibilities of adaptive behavior that nervous systems can achieve with their two most distinctive features: connectivity and timing.

Timing and Circuitry

The idea of "truth for the jellyfish" quoted at the beginning of this chapter would probably be taken as entirely facetious if not for its source. (T.S. Eliot was not notable for his rollicking sense of humor.) Jellyfish are not generally considered to have much in the way of epistemology. There is, however, more to it than even Eliot imagined, if we consider where epistemology originates. To the extent that we understand its origins, possession of a large, complex brain seems to be essential. Complexity in a brain has everything to do with circuitry and timing. The more brain you have, the greater your repertoire of circuits, and the more intricate and sophisticated the combinations of them that you can put together. For this reason, understanding brains requires analysis on multiple levels. The action is not in any individual cell, but rather in ensembles of cells.

In our brains, the number of possible ensembles is astronomical. Sorting among this number of possible activities, as well as making the most of its combinations in linking perceptions and actions, requires careful coordination and precise timing. Thus, the jellyfish has more to teach us about this vital and enormously complex subject than first meets the eye.

Further Reading

Gorman A.L.F., Hermann A., and Thomas M.V. 1981. Intracellular calcium and the control of neuronal pacemaker activity. *Fed. Proc.* **40:** 2233–2239.

Kristan W.B., Jr., Calabrese R.L., and Friesen W.O. 2005 Neuronal control of leech behavior. *Prog. Neurobiol.* **76:** 279–327.

Mackie G.O. 2004. Central neural circuitry in the jellyfish *Aglantha*. *NeuroSignals* **13:** 519.

CHAPTER 4

Modulation
The Spice of Neural Life

If "compression is the first grace of style,"
you have it. Contractility is a virtue
as modesty is a virtue.
It is not the acquisition of any one thing
that is able to adorn,
or the incidental quality that occurs
as a concomitant of something well said,
that we value in style,
but the principle that is hid:
in the absence of feet, "a method of conclusions";
"a knowledge of principles,"
in the curious phenomenon of your occipital horn.

To A Snail by Marianne Moore

The capabilities of invertebrates have traditionally been underestimated. Perhaps this is because they are not warm and fuzzy, or because they do not make very affectionate pets. Almost certainly, it owes something to the fact that the world as they find it is so different from ours that it takes some effort to put ourselves sufficiently in their place to devise tasks that will reveal their talents. For whatever reason, it has taken us an inordinately long time to realize that even the simplest animals have the capacity for modifying their behavior by adjusting the activities of their nervous systems. Perhaps this is a fundamental, inseparable property of nervous systems.

Snails in Trouble

Like most creatures in our competitive world, the otherwise peaceful sea snail lives under the threat of predators. Because *Aplysia* are not exactly equipped for a quick

Figure 4.1. The inking response of *Aplysia*.

escape response along the lines of an animal such as the jellyfish *Aglantha* (Chapter 3), they have evolved different responses to such threats. When pinched, bitten, or otherwise cut on their mantle by a spiny lobster, they will quickly close up their gill and siphon in a response that is not very different from the barnacle's shadow response (Chapter 2). In addition, they also eject a cloud of ink that surrounds them with a protective buffer (Fig. 4.1). This turbid cloud does more than just obscure the lobster's vision; it also contains a cocktail of unpleasant and confusing chemicals, some that taste foul to the lobster and others that stimulate the lobster's feeding response. The effect of stimulating the feeding response is to confuse the predator into thinking that it has already started eating, so that it ceases pursuit and commences its program of feeding movements before it has actually snared any prey.

As the sea snail escapes, it carries with it a souvenir of its dangerous encounter: It has now become hypersensitive, so that the next time anything touches it, the gill and siphon (Fig. 4.2) retract even more quickly. In other words, the reflex of closing up gill and siphon has been tuned up to require less of a stimulus. (The technical term for this is "once bitten, twice shy.") When recordings are made from the various neurons that perform the gill or siphon withdrawal response, it is clear that the harsh experience has increased the potency of neurotransmission at key synapses. Particularly important are the synapses from the sensory neurons that respond directly to touch, the synapses onto the motor neurons that induce contraction of the withdrawal muscles, and various cells (interneurons) in between (Fig. 4.3). The result is more contraction for less stimulation.

How does a synapse "know" it is supposed to modify its response after the animal is bitten? Because telepathy is not an option, there must be one or more signaling mechanisms, and because this is biology and not engineering, it turns out to be several mechanisms rather than one.

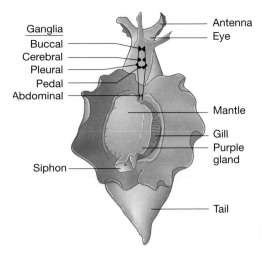

Ganglia
Buccal
Cerebral
Pleural
Pedal
Abdominal

Antenna
Eye

Mantle

Gill
Purple
gland

Siphon

Tail

Figure 4.2. Anatomy of *Aplysia* showing external structures and nervous system (ganglia).

Aplysia has a kind of alarm system within its nervous system, consisting of a set of neurons that project among all of the ganglia. Some of these neurons are activated when any part of the animal's body is touched, disturbed, or injured (Fig. 4.3). The more severe or long-lasting the disturbance, the more these "alarm" cells fire (an example of one of these cells is shown in Fig. 4.4). The neurotransmitter that is released by these alarm cells, serotonin, acts differently from the sorts of transmitters described previously (Chapters 2 and 3). Rather than opening ion channels coupled to a receptor for a fast response, serotonin activates receptors coupled to enzymes that configure slower chemical signaling cascades. The result is that the cell modifies

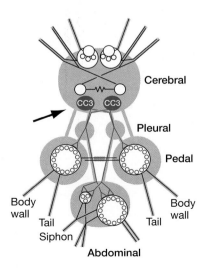

Cerebral

CC3   CC3

Pleural

Pedal

Body
wall

Body
wall

Tail

Tail

Siphon

Abdominal

Figure 4.3. The location of serotonin-containing CC3 cells (*arrow*) in the *Aplysia* nervous system. Shown are their projections to the pedal and abdominal ganglia, as well as the inputs from sensory neurons in the body wall, tail, and siphon to those ganglia. The cerebral ganglia receives sensory input from the neck, lips, antennae, and other head structures.

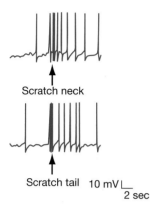

Scratch neck

Scratch tail   10 mV⌐
                   2 sec

Figure 4.4. Responses of serotonin-containing neuron CC3 to a scratch on the neck or tail. The firing rate of action potentials increases markedly upon stimulation and continues for several seconds.

its excitability over longer periods of time. This type of signaling is more akin to the rhodopsin cascade described in Chapter 2, but with an even longer-lasting effect. In addition to being released at synapses, some of the serotonin is also released diffusely in the nervous system, especially after a particularly noxious stimulus. This ensures that its effects will be widespread and long lasting.

One result of serotonin release is increased efficiency of synpatic release from the sensory neurons involved in *Aplysia*'s retraction response. There are serotonin receptors on these neurons, and when the transmitter is applied directly to them, the duration of each action potential increases (i.e., the spikes become broader). As a result, the nerve terminal is depolarized for a longer time and more transmitter is released. Postsynaptic depolarization in the motor neurons is correspondingly increased (Fig. 4.5). Depending on the strength of the touch or injury, this sensitized response can last from minutes to hours.

If one of the animal's sensory neurons is actually severed, for instance after a spiny lobster rips through the skin, the remaining end of that neuron becomes hyper-active in firing action potentials and remains in that state for some time. This hyper-excitability response is produced by a different mechanism than either of those described above, one that does not involve serotonin but instead appears to be an intracellular response to the injury itself.

Modulating Neuronal Activity

Serotonin is one of a class of transmitters known as "neuromodulators" that alter activity in the nervous system over relatively long time frames. It is best known to us through its effect on mood; a serotonin transporter in our brains is the target of the drug Prozac. Although neuromodulators act more slowly than ion channels, they often exert their effects by modifying ion channels.

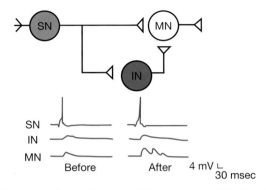

Figure 4.5. Simplified diagram of the circuit for gill retraction (*top*) in the abdominal ganglion. *Triangles* denote synapses. (*Left*) Stimulation of sensory neurons (SN) normally results in gill withdrawal by muscles driven by the motor neurons (MN). (*Right*) After a sensitizing experience (such as being attacked by a lobster), the synapses between the sensory neuron and the motor neuron, as well as between the interneuron (IN) and the motor neuron, are modified so that a larger response is produced in the motor neuron. The circuit actually consists of populations of sensory neurons, interneurons, and motor neurons.

Aplysia's sensory neurons have two different types of receptors on their surface for serotonin—one activated by low levels of stimulation and another activated by higher levels. After a mild touch, which causes a small amount of serotonin release, the first class of receptor binds the transmitter and changes shape, analogous to rhodopsin's shape change after absorbing light (see Chapter 2). In fact, serotonin receptors are in the same general family of proteins as rhodopsin. As with rhodopsin, the shape change allows it to associate with a G protein and to activate an enzyme (Fig. 4.6). The enzyme is adenyl cyclase (AC), which makes cylic AMP (cAMP). cAMP then binds to another enzyme, protein kinase A (PKA), and activates it by causing an inhibitory subunit to fall off. As a kinase, PKA is in the class of enzymes that phosphorylate other proteins. In the case here, active PKA phosphorylates one of the sensory neuron's potassium channels (K_s), which results in the channel's inactivation. The net effect of inactivating this class of potassium channel is that the next time an action potential is triggered in the neuron, repolarization will take longer. The slower repolarization broadens the spike and permits the entry of more calcium than previously would have occurred during one action potential; thus, more neurotransmitter will be released each time.

A second class of receptors on these neurons requires a longer exposure to serotonin before activating its coupled enzyme, phospholipase C (PLC). When active, PLC cleaves one of the lipids in the membrane to produce the intracellular signal diacylglycerol (DAG). DAG activates the enzyme protein kinase C (PKC), which phosphorylates and inactivates yet another class of potassium channels. This is a different class of potassium channel than those modified by PKA, and now the sensory neu-

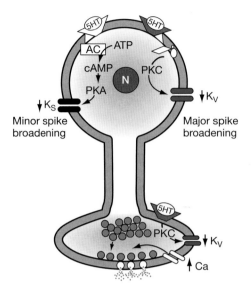

Figure 4.6. Kinases and channels participating in sensory neuron sensitization (see text). (*Top*) Cell body with nucleus (N); (*bottom*) synapse with vesicles. The two classes of serotonin receptor on the cell body are represented by *white* and *black trapezoids*, each linking to a different enzymatic cascade that modifies a different set of channels either in the cell body or in the synaptic terminal. (5HT) Serotonin; (AC) adenyl cyclase; (cAMP) cyclic AMP; (ATP) adenosine triphosphate; (PKA) protein kinase A; (PKC) protein kinase C; (K_S) potassium channel initially modified after brief serotonin stimulation; (K_V) potassium channel modified after longer exposure to serotonin stimulus.

rons are even slower to repolarize, release even more transmitter per spike, and activate the motor neurons of the retraction response even more easily.

There are three key steps to the regulation of this response: (1) the widespread release of serotonin after a stimulus to the skin, (2) the presence of receptors on the cells with two grades of response, and (3) the modification of two different kinds of potassium channel in the sensory neurons, each a target of a separate protein kinase activated by its respective receptor.

For *Aplysia*'s hyperexcitability response to a localized cut, which does not rely on serotonin, the localized response does not work through either the cAMP/PKA system or the DAG/PKC system. Rather, it operates through a less well-characterized signaling pathway that appears to involve yet another protein kinase, in this case, one regulated by cGMP.

Evolution of Neuromodulation

With the realization that an animal as simple as *Aplysia* exhibits experience-dependent modifications in its nervous system, it is natural to assume that such mechanisms must be very widespread. (If *Aplysia* does it, then anyone can do it!) Such a conclusion, however, turns out to be false. Even a limited survey of closely related sea snail species (Fig. 4.7) shows that some have it (*Aplysia*) and others either lack it or have only certain aspects of it (*Phyllaplysia* and *Dolabrifera*).

Certainty in evolutionary questions is always elusive, but one can infer possible

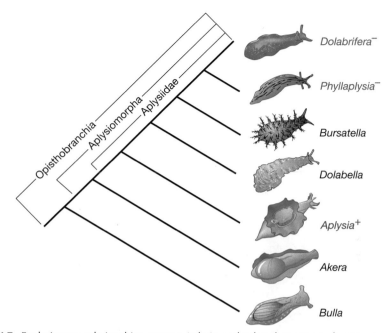

Figure 4.7. Evolutionary relationships among *Aplysia* and related species indicating presence (+) and absence (–) of sensitization. (*Left*) Taxonomic groupings: subclass Opisthobranchia, order Aplysiomorpha, and family Aplysiidae.

answers from the circumstantial evidence. Ecological characteristics of three of these species may have influenced whether modifiability was adaptive. *Aplysia*, which has the full capability, lives in the rocky bottom, near shore habitats along the Pacific coast where its predators include spiny lobsters, anemones, and various kinds of fish. *Phyllaplysia*, in contrast, is much smaller and lives on the blades of eel grass in the Pacific northwest, where it is likely to be much less subject to predation. *Dolabrifera*, living on the underside of boulders in tropical seas, may also be much less exposed to predation than *Aplysia*.

   If some of *Aplysia*'s cousins do not have the same responses to insult and injury, which components of the mechanism are they missing? At the level of physiology, there is the broadening of action potentials in sensory neurons due to the activity of serotonin ("spike broadening"). This response can be tested readily by squirting serotonin onto the sensory neuron to determine whether its action potential lasts longer. At the level of behavior, there are two aspects: "general sensitization," in which the animal responds to an initial insult (e.g., to the tail) by becoming generally skittish to stimuli all over, and "local hyperexcitability," in which direct injury to sensory neurons produces a hypersensitivity to subsequent stimuli that is confined to that side.

Dolabrifera fails to show an increased response of any type—no spike broadening, no local hyperexcitability, and no general sensitization (Fig. 4.8). *Phyllaplysia* exhibits spike broadening and local hyperexcitability, but no general sensitization. So far, these are the only species that have been tested for general sensitization. The other species all show some degree of local hyperexcitability, but they vary in their display of spike broadening.

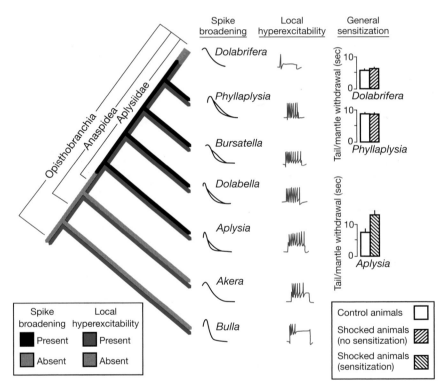

Figure 4.8. Phylogeny of serotonergic response and hyperexcitability. The distribution of spike broadening, local hyperexcitability, and general sensitization among six relatives of *Aplysia* is mapped onto their phylogenetic tree. Spike broadening is represented by a separation between the two *black traces* in the first column; local hyperexcitability is represented by multiple *spike traces* in the second column; general sensitization is indicated as a behavioral score in control versus stimulated (shocked) animals in the third column. *Black lines* in the phylogenetic tree depict evolution of spike broadening, and *dark blue lines* of increased local hyperexcitability. Increased local hyperexcitability was present in the ancestor of all species studied, but was lost in the lineage leading to *Dolabrifera*. Spike broadening was absent in the ancestral species, evolved in the aplysiid group (containing all but *Akera* and *Bulla*), and was then lost, along with spike broadening, in the lineage leading to *Dolabrifera*. General sensitization in *Aplysia* follows the presence of the neuromodulatory traits in this species, and the lack of sensitization in *Dolabrifera* is consistent with its lack of neuromodulatory traits. The lack of sensitization in *Phyllaplysia*, however, is at odds with the presence of the neuromodulatory traits in this species.

One would have thought that spike broadening and general sensitization would always go together, with one serving as the underlying mechanism for the other. *Phyllaplysia*, however, contradicts this expectation. It fails to show general sensitization, but its sensory neurons do respond to applied serotonin by broadening their spikes. How can that be?

A check of the neuroanatomy of these creatures showed that they all have serotonergic neurons in the right place, even *Dolabrifera* (Fig. 4.9). Furthermore, all of these cells are capable of releasing serotonin after tail stimulation (Fig. 4.10). The differences in release correlate to some degree with the species' behavior, but not completely. The paradoxical case is *Phyllaplysia*, which releases a fairly large amount on the ipsilateral side, and relatively more than *Aplysia* on the contralateral side, despite the fact that it has no sensitization whatsoever.

What appears to have occurred in the *Aplysia* lineage is a sequence of evolution-

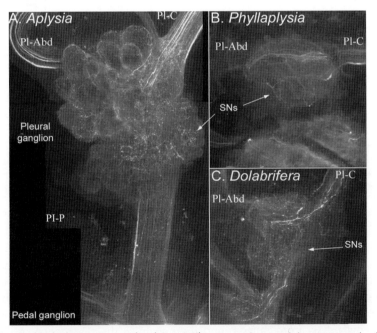

Figure 4.9. Confocal photomicrographs show similar serotonin-containing neurons in *Aplysia* (A), *Phyllaplysia* (B), and *Dolabrifera* (C). The orientation in all frames: pleural ganglion (*top*), pedal ganglion (*bottom*), cerebral-pleural nerve (Pl-C; *top right*), pleural-abdominal nerve (Pl-Abd; *top left*), and pleural-pedal nerve (**Pl-P**; *bottom*) (see Figs. 4.2 and 4.3). The three species show a similar pattern of serotonin-containing neurons: Numerous, brightly stained serotonin-containing cells are present in the cerebral-pleural nerve, which enters from the top right, and they branch over the cluster of tail sensory neurons (SNs; *arrows*) and/or travel to the abdominal ganglion through the pleural-abdominal nerve, which exits at the top left. Some serotonin-containing axons also travel toward the pedal ganglion through the pleural-pedal nerve, which exits at bottom. Scale bar, 100 μm.

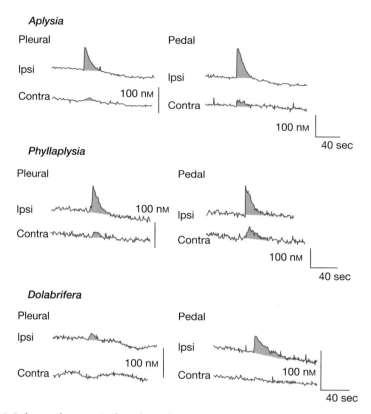

Figure 4.10. Release of serotonin from homologous neurons (see Fig. 4.9) after tail stimulation in *Aplysia* and its relatives. *Blue areas* under each trace represent the total amount of serotonin released on the same (ipsi) or opposite (contra) side as the tail stimulation from either of two ganglia, pleural or pedal (see Fig. 4.3). Scale bars at lower right for each group of traces indicate concentration of serotonin (nM) versus time (sec).

ary acquisition and loss of the response to serotonin by tail sensory neurons. The distribution of traits through the branches of the lineage of Ophistobranchia, the subclass containing these species (Fig. 4.8), suggests that this response was acquired before the divergence of *Aplysia* from *Akera*, then lost in two stages sometime after the divergence of *Phyllaplysia* from *Aplysia*, and then again in the divergence of *Dolabrifera* from *Phyllaplysia*. In the first of these loss steps (between *Aplysia* and *Phyllaplysia*), the sensory neurons respond to serotonin, which is still released from the neighboring neurons, but the behavioral response is gone. Presumably, *Phyllaplysia* are missing some component elsewhere in the circuitry. In the second of these steps (between *Phyllaplysia* and *Dolabrifera*), the sensory neurons no longer respond to serotonin, nor do they become locally hyperexcitable after injury. Exactly which components in the process have been lost or altered remains to be seen. When revealed, they will give us a small window into how modifications of the nervous system occurred during evolution. As is already clear from the known alterations, efficiency is not the driving force. "Useless" receptors may be left in place, perhaps to be revived at some later evolutionary epoch, or perhaps ultimately to be lost.

Neuromodulation and Specificity in the Nervous System

Neuromodulation does many things. As shown above, it can bias the responses of neurons after an unpleasant experience so that their reaction is amplified. In a similar vein, it can act as a reinforcer and mediator of reward in the process of learning. Some of the same transmitters, circuits, and behaviors described above are also involved in associative conditioning in *Aplysia*. We associate associative conditioning with Pavlov's dogs, who made history with their response to the following sequence of stimuli: After a neutral stimulus (a bell) was paired with a salient stimulus (tasty raw meat), the bell by itself took on the attractiveness of the meat, so that the bell alone induced the dogs to salivate. In the case of *Aplysia*, the neutral stimulus is a mild touch to the animal's siphon, whereas a harsh shock to the animal's tail is the salient stimulus.

Some neuromodulators set overall levels of arousal and depression in the nervous system, as in the well-publicized antidepressant action of Prozac. By blocking a serotonin transporter protein in our brains, Prozac increases serotonin levels at synapses. Like the barnacle's histamine transporter protein (see Chapter 2), the serotonin transporter pumps the neurotransmitter back into neurons after its release. When blocked, transmitter persists for longer in the synapse to activate receptors on the postsynaptic cell. Another example is the drug methamphetamine, which acts as a potent stimulant, causing hyperactivity and suppressing sleep. It blocks the transporter protein for another neuromodulator, dopamine. Such effects on neuromodulation create a system-wide bias in the activity of a great many neurons in the brain, which can globally affect an individual's behavior.

Up to this point, we have been operating on the assumption that specificity in the output of the nervous system is a function of timing and circuitry (see Chapter 3). Neuromodulation adds a third dimension by allowing patterns of neural activity to adapt to new conditions.

Judging from their ubiquitous presence, even in the most primitive animals, neuromodulators appear to go back to the very origins of nervous systems. Serotonin is found in Planaria (flatworms) and Cnidarians (e.g., jellyfish). Neuropeptides, an enormous class of neuromodulators, are also found in all of these organisms. In fact, serotonin and neuropeptides are present in sponges, which lack full-fledged nervous systems but do have excitable epithelial cells. To the extent that the evolutionary record is reflected in what has survived, it tells us that to have a nervous system is to have a neuromodulatory system. Perhaps you cannot have one without the other. Why might this be true?

Variability in the Nervous System: Wasteful or Useful?

As judged by engineering standards, nervous systems are noisy. The signals in neurons are often not reproducible, as is described below. Synaptic release can vary tremendously, depending on the strength of stimuli to a cell (see Chapter 3) as well as on the cell's immediate past history, as in *Aplysia*'s sensitization or Prozac's effects on humans. These effects are magnified many times over by the fact that neurons are highly interconnected in a dense network. In the only animal for which a full wiring diagram of the nervous system has been worked out, the nematode *Caenorhabditis elegans*, it has been possible to estimate the degree of connectedness. Theoretically, a sensory stimulus to four of the animal's sensory neurons would reach 6% of the animal's approximately 300 neurons directly in one step (i.e., through their first synaptic connections) and 50% of the nervous system in two steps. That is in theory, but what about reality?

Using an optical technology that allows hundreds of neurons to be monitored for their electrical activity at the same time, the reproducibility of *Aplysia*'s gill withdrawal response was recorded during repeated mild stimuli to the siphon (Fig. 4.11). There was significant variation in the number of cells, the identity of cells, and the strength of response of cells from one trial to the next. The number of cells responding is in the hundreds. This rather sobering result should serve to remind us that the simplifying circuit diagrams that we are so fond of drawing (e.g., Fig. 4.5) are indeed simplifying and therefore more useful to us for their heuristic value than for their accuracy.

These *Aplysia* findings reinforce the theoretical prediction obtained in the nematode: It is almost impossible to insulate the activities of neurons in one region from those in another. This means that some other strategy, aside from insulation, must be exerting itself. If the intrinsic variability in the nervous system cannot be reined in, then perhaps it is exploited. We have seen in previous chapters that evolved systems rely on degeneracy as a means of buffering against fluctuations and of ensuring the reliability of the ultimate outputs (see Chapters 1 and 2).

In reality, nervous systems are potentially chaotic places. If one estimates the actual extent of connectivity in a system such as *Aplysia*'s gill withdrawal response, the potentially relevant circuitry is huge and tangled (Fig. 4.12), with many hundreds or perhaps thousands of potential pathways intervening between sensory neurons and motor neurons. The response is distributed widely in the nervous system. Granting that not all such pathways are necessarily productive or consequential in their output, if they all exist and have some probability of occurring, and if many of them produce at least some degree of consequential output, how can the animal sort out the useful from the wasteful?

We have seen in this chapter that neuromodulation serves to bias the activities of neurons. What do these biases do to variability? It appears that they help rein it in, as shown in another experiment monitoring a very large number of cells in the *Aplysia* nervous system (see Fig. 4.13), in which the entire abdominal ganglion was

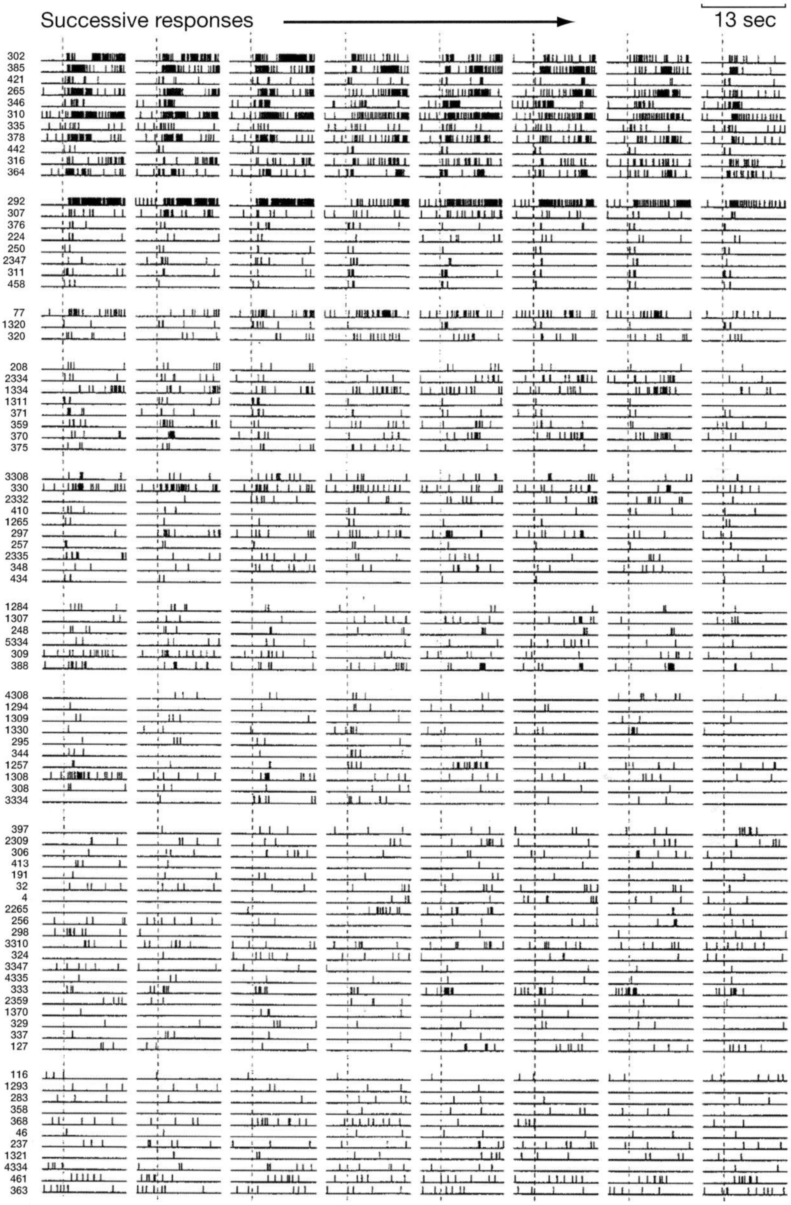

Figure 4.11. Trial-to-trial variation in the number of neurons responding and the intensity of their responses in the abdominal ganglion of *Aplysia* after a mild stimulus to the tail. Eight successive presentations of the stimulus are shown (*left to right*) and neurons are grouped (*top to bottom*) according to the intensity of their response. Numbers at left represent individual cells and each tick represents an action potential.

Sensory
neurons

Interneurons

Gill
motor neurons

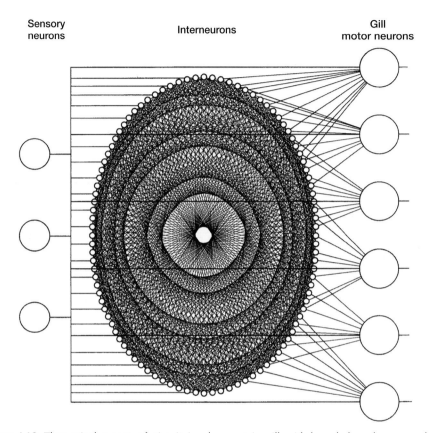

Figure 4.12. Theoretical extent of circuit involvement in gill withdrawal, based on population recordings from *Aplysia* (as in Fig. 4.11) and calculations from *Caenorhabditis elegans* estimating that a sensory stimulus to a subset of the animal's sensory neurons would reach 6% of all other neurons directly by one synaptic connection and 50% of the neurons after a second synaptic connection. The opaqueness of the oval-shaped area in the diagram reflects the very large number of connections among neurons that would potentially be involved in the response.

monitored before and after sensitization. Clearly not all variability is suppressed, but there has been a modulation of the cells involved, consisting of a small increase in the number of responders (from 103 to 113 neurons) but a reduction in the response strength of most neurons. Why would there not simply be a wholesale increase in the number and strength of cells that respond to a process that enhances responsiveness? This is partly because some of the enhanced cells are inhibitory and thus dampen the activity of other cells, which helps to focus the system and suppress the tendency of highly interconnected networks to excite themselves out of control. Out-of-control excitation is what occurs in epilepsy. On the other hand, the system response (Figs. 4.11 and 4.13) is still neither highly confined nor highly reproducible.

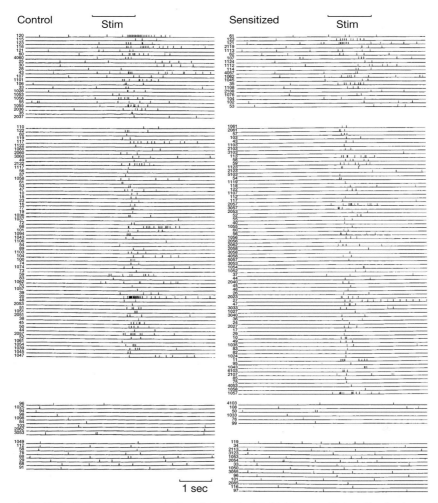

Figure 4.13. Gill withdrawal response before (*left*) and after (*right*) a sensitizing stimulus (recordings as in Fig. 4.11). Although the behavioral state is heightened after sensitization, the actual number of neurons activated is reduced as compared to the unsensitized (control) animal.

Decision-making and Variability

Another explanation of how the nervous system sorts out the useful from the wasteful may be found in the way ganglia in the leech make "decisions." The leech, as discussed in Chapter 3, has an elaborate circuitry to generate rhythmic firing of neurons as part of its swimming behavior. These neurons are in no sense dedicated to swimming. They can also be recruited into different ensembles that give rise to other behaviors, one of which is crawling. In fact, when an optical recording technique is used (similar to the one just described for *Aplysia*), all of the approximately 140 visible cells in the ganglion are active in both swimming and crawling. The behaviors cannot be distinguished from one another simply by which cells are involved. The patterns of their activity, however, such as firing frequency and the phase of firing between cells, form the very basis of which behavior is being produced. Once each behavior has kicked in, there is a clear distinction in the patterns of activity between the two, although it is not an exact match, cell for cell, each time.

Initially, however, no clear distinction is apparent. This can be seen with a touch stimulus to the leech's body wall that will elicit either a swimming or crawling response from the nervous system with roughly equal probability. In this sense, the leech's neurons "choose" between the two behaviors. Initially, there is actually no difference in the cells' response pattern to the stimulus. After an initial period of similarity, the responses eventually diverge from one another (Fig. 4.14). When the ensemble of their activities is analyzed in an effort to find the first indicator of which direction will be taken, there is no single cell that reliably predicts the choice. Instead, the statistically correlated activity patterns from a group of cells predict the choice, and stimulation of at least one of these cells can bias the choice. These findings suggest that decision-making is a property of the network. An individual cell's activity can influence the probability of one choice versus the other, but the choice

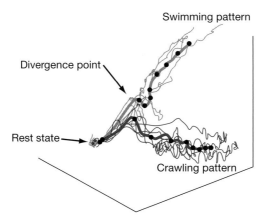

Figure 4.14. Representation of patterns of activity of 140 neurons in a leech ganglion beginning with a touch stimulus to the body ("Rest state") to the final differentiated pattern of swimming (*gray*) versus crawling (*blue*). Each axis represents a characteristic statistical feature (principal component) of the population. The lines show how the population activity evolves at successive time points (*left to right*), at first indistinguishable and then diverging. (*Narrow lines*) Individual experiments, no two of which are identical; (*thick lines with points*) averages for each group. Note that some trajectories begin to migrate toward one pattern after the "Divergence point" and then make a sharp turn toward the other.

occurs at the level of the network. The behavior of the leech ganglion also shows that there are multiple (i.e., degenerate) ways in which the circuit choices can be achieved.

Is the leech's choice entirely random? For a behavior to be adaptive for the animal, there must be ways of influencing the response. We have already seen that neuromodulators are effective in introducing such biases, and the leech is no exception. Serotonin and other neuromodulatory transmitters act at various points in the leech nervous system to nudge the response one way or the other.

Neuromodulation and the Cold, Cruel World

So maybe neuromodulation, in conjunction with the properties of circuitry and timing, allows nervous systems to avoid chaos without having to give up their intrinsic variability. But why is it so important to maintain variability? It is because the world is a harsh and variable place. Organisms are constantly coming up against a changing set of circumstances and their senses are continually bombarded with a massive and constantly fluctuating jumble of stimuli. Neuromodulatory systems go a long way toward making nervous systems able to cope with these fluctuations. They allow all of us creatures (invertebrate or otherwise) to deal effectively with the world as we find it.

Further Reading

Briggman K.L., Abarbanel H.D., and Kristan W.B. 2006. From crawling to cognition: Analyzing the dynamical interactions among populations of neurons. *Curr. Opin. Neurobiol.* **16:** 135–144.

Byrne J.H. and Kandel E.R. 1996. Presynaptic facilitation revisited: State and time dependence. *J. Neurosci.* **16:** 425–435.

Wright W.G. 2000. Neuronal and behavioral plasticity in evolution: Experiments in a model lineage. *Biosciences* **50:** 883–894.

Wu J.Y., Cohen L.B., and Falk C.X. 1994. Neuronal activity during different behaviors in *Aplysia*: A distributed organization? *Science* **63:** 820–823.

CHAPTER 5

An Internal Wake-up Call

Here come real stars to fill the upper skies,
And here on earth come emulating flies
That though they never equal stars in size
 (And they never were really stars at heart),
Achieve at times a very star-like start.
Only, of course, they can't sustain the part.

Fireflies in the Garden by Robert Frost

Just before dawn in the vineyard, the fruit flies are starting to stir. When the first faint light appears, they will be up and out to forage and get their business done before the sun gets too hot. Even before the first light, they become restless as their nervous systems gear them up for action. Like virtually all living things, they have an internal clock that sets their "circadian rhythm," the time during the 24-hour cycle when they will be active. This clock is set by the rising and setting of the sun (or the turning on and off of lights in a laboratory incubator), but it is not merely registering lights-on and lights-off. Instead, it prods the fly to become active in anticipation both of the first light and of lights-off in the evening (Fig. 5.1). If fruit flies jet-setted around the world the way that we do, they too would experience jet lag. More dramatic proof of the flies' ability to keep time appears when they are kept under conditions of constant darkness, as happens to Icelandic fruit flies during winter months. Then, if they were surviving out in the cold, they would continue to cycle through active and inactive periods on nearly the same schedule as if they were experiencing day and night. This is what flies do when they are kept in constantly dark (but comfortably warm) incubators (Fig. 5.2). The average length of these virtual day/night cycles is right around 24 hours.

Where is the fly's clock? The surprising answer is that it is everywhere. That is, many (perhaps all) of the cells in the fly have the molecular machinery for keeping 24-hour time. There are, however, groups of neurons in the brain that coordinate the clock's effects on behavior (Fig. 5.3). These cells contain the machinery for keeping

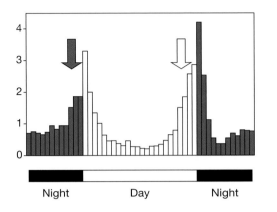

Figure 5.1. Activity pattern throughout day and night of a 24-hr period, showing (*arrows*) anticipatory activity before lights-on and lights-off. The 24-hr period is indicated by the *horizontal black* and *white bars*, and each vertical bar of the histogram represents the fly's movement during that hour.

24-hour time and they also form a network of interactions that coordinates the setting of the clock to the cycle of daylight and nightfall in which the fly finds itself. The timing is then communicated to the rest of the brain, where different behaviors will occur preferentially at different times of day. For the fly to anticipate daybreak and nightfall, its internal clock must be correctly set to the appearance and disappearance of daylight. And in all parts of the world except the Equator, this changes constantly over the course of the year, so continual resetting is necessary.

Setting the Clock

The fruit fly uses three systems for detecting light and setting the clock. The principal one, not surprisingly, is its external eyes, which consist of the two large compound eyes on either side of its head as well as a trio of small photoreceptor structures on the top of its head called ocelli (Fig. 5.3). These all have standard, invertebrate, photoreceptor cells containing rhodopsin and its attendant phototransduction cascade (see Chapter 2).

Through a series of synapses and intermediary neurons in the optic lobes that are the visual centers of the fly's brain, the electrical signal instigated by the first light of the morning reaches the outermost branches of a cluster of clock neurons known as

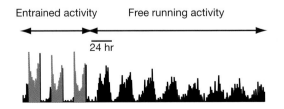

Figure 5.2. Activity rhythm during 3 days of entrainment to a light–dark (*blue/black*) cycle followed by 8 days in constant darkness. Bar, 24 hr.

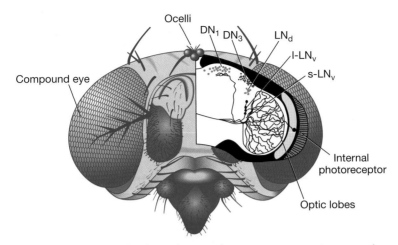

Figure 5.3. Frontal view of the fly's brain showing photoreceptor organs (compound eye, ocelli, and internal photoreceptor), optic lobes, clusters of dorsal (DNs) and lateral (LNs) clock neurons, and the branching pattern of one set of LNs.

the ventral lateral neurons (LN_v, Fig. 5.4). These outer branches, called dendrites, radiate widely throughout the optic lobes, allowing them to pick up light responses from any part of the visual field.

Inside a group of small lateral neurons (s-LN_v), there is also a light-sensing molecule called Cryptochrome (Cry), an ancient protein that is conserved among animals, plants, and bacteria, which uses a flavin molecule for sensing light in the blue part of the spectrum. (Recall that flavins are also used as light sensors in sponge larvae; see

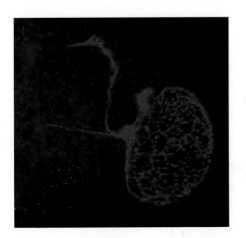

Figure 5.4. LN_v cells and their dendrites.

Chapter 2.) Cry-containing s-LN$_v$ cells serve as a second system of light detection for clock setting and, because the fly's head is translucent, some light reaches these cells directly. More importantly, Cry is also an essential component of the clock-setting mechanism: It initiates a series of reactions that modifies the intrinsic timing and pacemaking function in these cells.

When Cryptochrome absorbs light, it changes conformation, allowing it to bind to a protein called Timeless (Tim, so named because mutant flies that fail to make the protein have a defective clock). Tim is an intrinsic component of the clock machinery. Complexing with Cry makes Tim ripe for degradation and a decrease in Tim levels resets the clock. As if to reinforce the fly's sensitivity to changes in the time of daybreak, the levels of Cry accumulate in the dark and decline in the light. This fluctuation is controlled by the light–dark cycle and helps to ensure that the cell is primed to respond to first light.

A third system for registering lights-on versus lights-off and entraining the clock is a small photoreceptor organ found just inside the retina at the outer margin of the optic lobes (Fig. 5.3). This simple structure also contains rhodopsin and is carried over through metamorphosis to the adult from the larva, where it may serve as the primary neuronal time-setting organ. Any one of the three systems is sufficient to elicit circadian rhythms, as shown by analysis of mutants that delete one or more of them. The little internal photoreceptor organ is the weakest in this regard, but it can nonetheless contribute to entrainment. In other words, the fly is well covered (i.e., degenerate) for this function.

The Clockworks

Circadian rhythms have been recognized for many years and some of the experiments designed to reveal and probe them have been heroic, such as submerging people in caves or diving bells for weeks with no clues to indicate the normal cycling of day and night. Cracking open the problem of the clock's mechanism, however, was less the result of heroics than of some simple, but clever, experiments. The first step was the isolation of a series of mutant strains of the fruit fly *Drosophila melanogaster* that appeared to have different free-running clocks.

An animal's free-running clock is measured by placing it in constant darkness after it has been entrained to a normal day–night cycle. The periodicity of the rhythm is determined from the average cycle of locomotor activity calculated over several days (Fig. 5.2), and symbolized by the Greek letter τ. In normal flies, $\tau = 24$ hours. The first of the fly mutants isolated had a τ of 19 hours. The second had a τ of 28 hours and the third had no discernible periodicity, that is, it was arrhythmic. The idea that these mutants might have something to do with the fundamental clock mechanism was suggested by the fact that each one affected the whole range of circadian controlled behaviors known at the time (locomotor activity, emergence of adult flies from metamorphosis, and egg laying). In other words, the mutations were not merely altering some specific aspect necessary for the output of a single behavior. Significantly, all three mutants turned out to be alleles of the same gene. This meant that the gene (dubbed *period*) was likely to be a fundamental component of the clock itself.

With the application of molecular techniques and the isolation of additional mutants (the second of which was the *timeless* mutant mentioned earlier as well as *doubletime, cycle, Clock,* and several others including *cryptochrome*), a mechanism for the clock began to emerge. The cellular machinery for 24-hour rhythmicity is a cycle of transcription of the *period* and *timeless* genes, translation of their mRNAs, and shuttling of these protein products (Per and Tim) back to the nucleus, finally resulting in a shutdown of *period* and *timeless* transcription by the repressive activity of the Per and Tim proteins on their own transcription (Fig. 5.5). As the Per and Tim proteins degrade and decrease back to a low level, the repression abates and their transcription begins once again; the cycle comes full circle. (This degradation of Per and Tim proteins is a distinct process from the Cry-induced degradation of Tim described earlier.)

The other two core components of the clock's molecular mechanism are Cycle (Cyc) and Clock (Clk), which are the proteins that activate transcription of the *period* and *timeless* genes, and with which Per and Tim complex to repress their own tran-

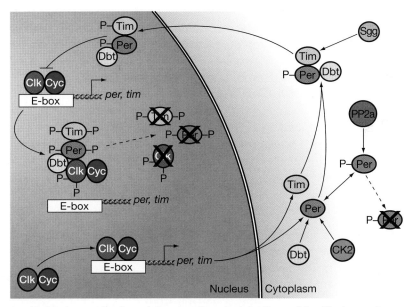

Figure 5.5. Molecular mechanism of the circadian clock in *Drosophila*. Clock proteins are Period (Per), Timeless (Tim), Doubletime (Dbt), Clock (Clk), Cycle (cyc), Shaggy (Sgg), Casein kinase 2 (CK2), and Protein phosphatase2a (PP2a); see text for explanation. Protein kinases Dbt and CK2 phosphorylate the Per protein, Sgg phosphorylates the Tim protein, and PP2a dephosphorylates the Per protein, all of which regulate the stability and localization of Per and Tim. Phosphorylation is shown by the (–P) attached to a protein. The E-box represents the sequence of DNA upstream of the *period* and *timeless* genes to which the Clk/cyc complex binds to activate transcription. Crossed-out proteins indicate degradation.

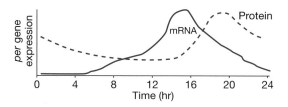

Figure 5.6. Cycling of Per mRNA and protein over a 24-hr period.

scription. Such feedback repression would not ordinarily produce a 24-hour cycle without the introduction of time delays into the cycle; otherwise, it would just reach a steady state and grind to a halt. Time delays occur between transcription and translation and then again in the shuttling of the proteins back into the nucleus. As a result, levels of *period* and *timeless* mRNA cycle in one phase and peak early in the night, whereas levels of Per and Tim protein cycle in another phase and peak late in the night (Fig. 5.6). Several protein kinases (Fig. 5.5) also take part in the process by acting to either regulate the degradation of Per and Tim or enhance their efficacy in repressing transcription.

Biochemically pleasing as this picture is, it is not the whole story. There is a second transcriptional loop of self-repression that affects the *Clock* gene. In this loop, Clk protein drives expression of two other rhythm-related genes, *vrille* and *Pdp1ε*, whose protein products then regulate transcription of the *Clock* gene.

This core mechanism makes the process of clock resetting somewhat clearer. Cry proteins would have built up during darkness, so an earlier sunrise would cause an earlier Cry-induced degradation of Tim, relieving its transcriptional repression and allowing the Clk and Cyc proteins to go back to transcribing the *per* and *tim* genes slightly earlier. Thus, rhythmic behavior changes its phase with the changing dawn.

Translating Time into Activity

Keeping and resetting time within the s-LN$_v$ cells is all very well, but if it does not get out to the rest of the nervous system, then it does not make any difference. An important element in the link between timekeeping and electrical activity in the brain is a neurotransmitter found in the s-LN$_v$ cells: a neuropeptide known as Pdf for pigment-dispersing factor. (Its name is derived from the fact that it was originally identified in fiddler crabs where it influences the movement of pigment in their retinal and epithelial cells.) A *Pdf* mutant strain in the fruit fly that lacks the neuropeptide produces a defect in rhythmic locomotor activity; specifically, mutant *Pdf⁻* flies lack the normal anticipation of morning light (Fig. 5.7). When the receptor (PdfR) for this neuropeptide (Fig. 5.8) is eliminated in a *PdfR* mutant, a similar loss of the morning anticipation peak occurs.

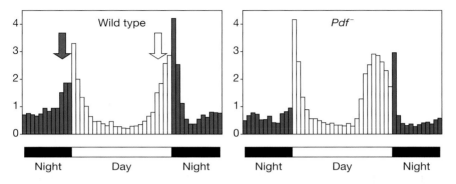

Figure 5.7. Morning activity peak in normal *(left)* and mutant *Pdf⁻* *(right)* flies. The 24-hr period is indicated by the *horizontal black* and *white bars*, and each *vertical bar* of the histogram represents the fly's movement during that hour.

Unifying Rhythmic Activity in the Nervous System

The loss of anticipation of the first morning light is not the only effect of lacking Pdf. These flies also gradually become arrhythmic over several days in constant darkness (Fig. 5.8). On closer inspection, Pdf is required to keep the various clock cells in the brain (see Fig. 5.3) synchronized with one another. This is seen by monitoring the cycling of Per in these cells. These cells all appear rhythmic in terms of their molecular cycling, but they all seem to be in different phases (i.e., starting the cycle at different times).

Not only do the Pdf-containing cells keep all of the other clock cells in line, they also coordinate timing of the anticipation of evening by signaling to another set of circadian pacemaker cells, called "evening" cells. (For symmetry, the Pdf-containing cells are also referred to as "morning" cells, see Fig. 5.9.) The evening cells can be

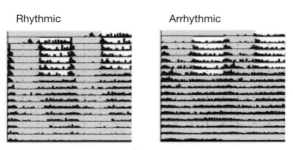

Figure 5.8. Rhythmic wild-type flies versus arrhythmic *Pdf⁻* mutant flies in constant darkness after 12 days of entrainment. Daylight periods are shown in *white* and *dark* periods are shaded *gray*. Traces are continuous activity records, 48 hr on each line, where *black ticks* indicate activity periods. Wild-type flies maintain their rhythmicity over many days in constant darkness, whereas *Pdf⁻* mutant flies start to become arrhythmic after several days.

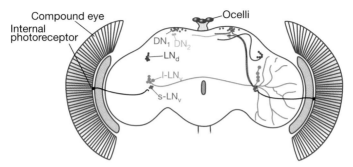

Figure 5.9. Morning (LN$_d$) and evening (s-LN$_v$) cells in the brain and their relationship to other clock neurons and visual organs.

eliminated genetically by directing the expression of a cell-killing gene specifically to those cells. Such flies lose the anticipatory evening peak of activity.

When morning cells reset their clock phase, for example, as the days lengthen in the spring, this is communicated to evening cells so that anticipation is present at the correct relative time. This can be shown by speeding up the internal clock in either of these two groups of cells, morning or evening, and asking whether one group directs the timing of the other group. The clock can be sped up by a bit of genetic engineering in which one of the clock-related protein kinases, Sgg (see Fig. 5.5), is expressed at an abnormally high level in one or the other cell group. Sgg overexpression produces a clock where $\tau = 21$ hours in constant darkness. When Sgg is selectively expressed in the morning cells, the evening cells follow suit and also speed up. The resetting only goes one way, however, as seen in the contrasting situation in which Sgg is selectively expressed in evening cells: No change in the anticipation of either evening or morning occurs.

Making Behavior Rhythmic

All of this still accounts for only part of the story. Light comes into the eye, stimulates the Pdf-containing (i.e., morning) cells, and allows them to set both their internal clock and the evening cells. But this is still a long way from orchestrating behavior on a circadian schedule. Do the clusters of circadian pacemaker cells do anything more than coordinate with one another? No doubt, they also signal elsewhere in the brain, given their extensive branches, but do they also have direct control over rhythmic behaviors that arise from specific parts of the brain? Or is it sufficient for them to be a whole brain metronome that tells local rhythm-controlled genes when to set the responses of cells? Upward of 100 genes show a circadian pattern of expression, suggesting that there is a circadian character to the differential gene expression that distinguishes subsets of neurons from one another in general. The brain's overall clock must somehow regulate the local transcription of these "output" genes. In this sense, the circadian system is another kind of modulator in the nervous system, complementary to neuromodulatory transmitters (see Chapter 4). Like neuromodulators, it is a nonlocal factor that interacts with local specific systems to modify activity in the nervous system. This interplay between general and local influences is a motif that appears repeatedly in the brain.

Sleep: More Than the Absence of Activity

Before their internal alarm clocks stir in the dark before dawn, why are fruit flies so quiet? It is not that they always stay put in the dark, because when experimentally subjected to constant darkness, they continue to be active during the period that would have been daytime. Instead, it seems that they are sleeping.

Flies do not snore, curl up in the corner, or close any of the 1400 ommatidia of their compound eyes. They just stand still (which is not adequate to qualify as sleep). In addition to being quiet, they also are less responsive to stimuli such as light or vibration, and if forced to become active during the nighttime when they would usually be quiet (i.e., become sleep deprived), they will recover some of the lost sleep during the next day. These are characteristics of sleep.

While a fly is sleeping, its brain becomes less active—not silent, as it does when the fly is completely anesthetized, but neural activity decreases. This can be seen in electrical recordings from a fly's brain (Fig. 5.10). (In contrast to the kinds of record-

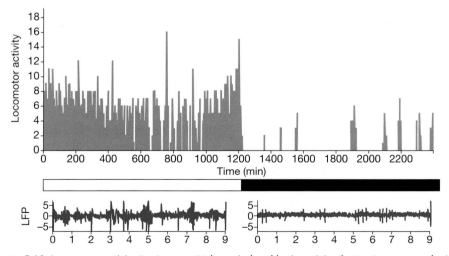

Figure 5.10. Locomotor activity (*top*) over a 24-hr period and brain activity (*bottom*) over several minutes in wakefulness (*left*) versus sleep (*right*). LFP (local field potential; see text) is the aggregate neural activity recorded from a region of the brain. The brain is quieter during sleep than during wakefulness.

ings shown in earlier chapters, Fig. 5.10 does not reflect what is going on in a single cell, but rather the electrical activity recorded extracellularly in one local region. These "local field potentials," or LFPs, represent the aggregate activity of all of the cells in the neighborhood of the electrode.)

Sleep is under the control of the circadian system, as can be clearly seen in the day–night difference in locomotor activity. Mutants of the *period* gene that have no rhythm in constant darkness still sleep, but not on a schedule. They sleep intermittently for short stretches through the 24-hour period. Their response to sleep deprivation, however, is intact: They recover the lost sleep when the deprivation ceases. What sleep does for the brain and why it is necessary remain open questions.

Evolution of Circadian Systems

Circadian rhythms are exceedingly ancient. Ubiquitous in the animal and plant worlds, they have also been found (and studied) in fungi, protozoans, and cyanobacteria (among the most primitive of all bacteria). The clock mechanism in mammals uses a set of genes that are homologous to those in the fly (Fig. 5.11). One does not find homologs of Per, Tim, and the others in fungi and cyanobacteria, but these organisms do have analogous mechanisms of transcriptional feedback. The principal rhythms that can be detected in cyanobacteria, in fact, affect the temporal pattern of expression for most of the cells' genes, which results in daily oscillations in nitrogen fixation, metabolism, and cell division. Clearly, an organism does not need a nervous system in order to have circadian rhythms.

The antiquity of the process is further suggested by the fact that the oldest cellular fossils, found in 3500-million-year-old rock, resemble an extant species of cyanobacteria called *Oscillatoria* (Fig. 5.12). Moreover, when the age of the primordial cyanobacterial rhythm gene is estimated from DNA sequence comparisons, an age of 3800–3500 million years is obtained. The Earth is 4500 million years old and it was not solid or cool enough for any life duing its first 500 million years. This suggests that synchronizing to the rising and setting of the sun was one of the initial adaptations in the history of life. τ may have been very different back then, because the Earth was spinning much faster, perhaps one revolution every 8 hours! When nervous systems evolved onto the scene, they took up the task of rhythm regulation, which was further elaborated over their increasingly complex networks.

Microevolution of Circadian Rhythms

Hints as to how rhythm genes actually evolve have been gleaned from looking at how they vary in today's world, which offers glimpses into the processes of selection and adaptation. The DNA sequence of the *period* gene varies considerably in populations of *D. melanogaster*. One portion of the gene that shows particular variation is a repet-

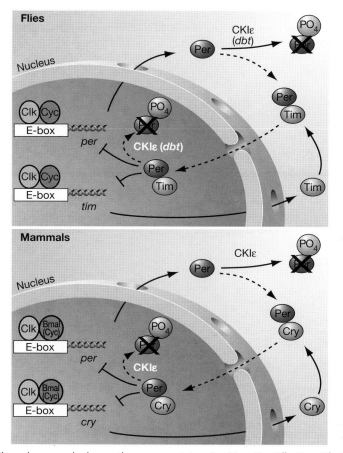

Figure 5.11. Fly and mouse clocks use the same proteins: Per, Tim, Cry, Clk, Cyc, Dbt (CKI), and Sgg in the fly all have mouse counterparts that are used in a similar, but not identical, mechanism. See Fig. 5.5 for explanation of symbols.

itive stretch of a pair of amino acids (threonine and glycine; Fig. 5.13). When populations from various geographical locations are sampled, it turns out that the distribution of these variants follows a north–south pattern that correlates with differences in climate (Fig. 5.14). This "latitudinal cline," as such distributions are known, suggests that there has been selection for certain alleles in colder climates. Can this be proven?

Selection is not easy to prove, but not impossible either. By examining the variation in DNA sequence on either side of the *period* gene, one can obtain a pretty good idea whether selection is occurring and, if so, what kind of selection. Three kinds of selection are possible: balancing (favoring more than one variant), directional (favoring one variant only), or neutral (no selection). It turns out that the *peri-*

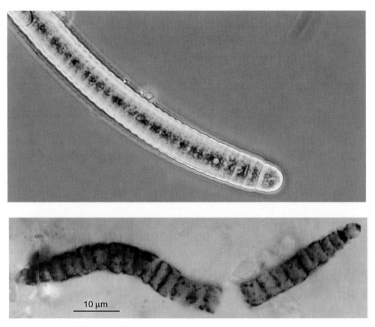

Figure 5.12. *Oscillatoria* (*top*) and its ancient forebear (*bottom*).

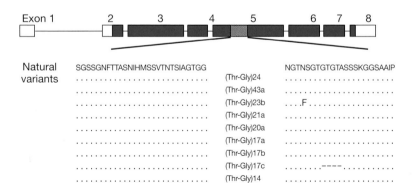

Figure 5.13. Threonine-glycine (Thr-Gly) repeat variants in the *period* gene. (*Top*) Structure of the *period* gene; *blue boxes* represent protein-coding sequences. (*Bottom*) Expansion of sequence in the light blue portion showing Thr-Gly variants found in natural populations of *Drosophila melanogaster*. Portions of the protein sequence flanking the Thr-Gly repeat (*dotted lines*) show almost no variation. The protein sequence is shown using single-letter amino acid symbols. (..F..) Amino acid substitution in the (Thr-Gly)23b variant; (----) four-amino-acid deletion in the (Thr-Gly)17c variant.

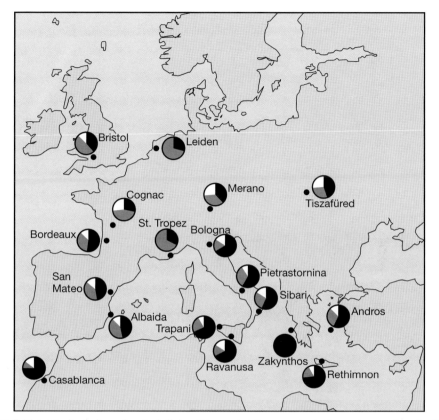

Figure 5.14. Latitudinal cline of *period* alleles. (*Pie graphs*) Frequencies in the various regions for the alleles (Thr-Gly)20a (*gray*), (Thr-Gly)17c (*black*), and all others (*white*). The (Thr-Gly)20a allele, which is more prevalent in the north, is more robust in its temperature compensation.

od gene has a DNA profile that is consistent with balancing selection, so the next step is to find out what the selective factor might be. Because temperature is the principal determinant in the apparent latitudinal cline, an obvious candidate is the clock's ability to compensate for temperature fluctuations. Such fluctuations would be more severe in northern climates than in southern, and could be particularly troublesome for fruit flies, who do not regulate their body temperature.

One of the more intriguing features of the circadian clock is its ability to keep accurate time over a wide range of temperatures, flying in the face of the usual temperature dependence of chemical reactions. The latitudinal *period* variants do, in fact, possess different abilities to compensate for temperature shifts (Fig. 5.15). The predominant northern-climate allele and southern-climate allele have each been geneti-

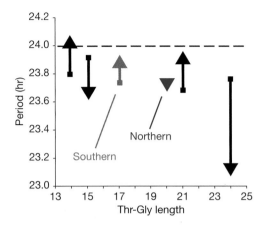

Figure 5.15. Temperature compensation by natural *period* Thr-Gly variants. The "northern" allele, (Thr-Gly)20, compensates better than the "southern" allele, (Thr-Gly)17. A 24-hour period in constant darkness was measured before and after shifting temperature and plotted versus Thr-Gly length. (*Squares*) Circadian period at 18°C; (*arrowheads*) at 29°C.

cally engineered into the standard laboratory strain of flies, so that the only difference is in that gene. In the resulting flies, the northern allele shows almost perfect temperature compensation, but with a period 15 minutes shorter than 24 hours. The southern allele, in contrast, shifts its τ by nearly 15 minutes with a temperature change, but it is closer to 24 hours at the warm temperature. Without knowing how the threonine-glycine repeat actually does this biochemically, the findings certainly suggest an adaptive advantage of the northern allele in its colder and more thermally variable environment. The advantage of the southern allele, on the other hand, is that at warmer temperatures, it beats at exactly 24 hours, in perfect tune with its cycling environment.

A Global Behavior

The conservation of circadian mechanisms across essentially all of phylogeny has been one of the great surprises of contemporary studies of molecular biology and behavior. Undoubtedly, such a degree of universality is related to the antiquity of the process. As in the history of channels, rhythms provide another example of the co-optation by nervous systems of preexisting mechanisms, which are then elaborated into more sophisticated forms.

Further Reading

Bell-Pedersen D., Cassone V.M., Earnest D.J., Golden S.S., Hardin P.E., Thomas T.L., and Zoran M.J. 2005. Circadian rhythms from multiple oscillators: Lessons from diverse organisms. *Nat. Rev. Genet.* **6:** 544–556.

Chang D.C. 2006. Neural circuits underlying circadian behavior in *Drosophila melanogaster. Behav. Processes* **71:** 211–225.

Hall J.C. 2003. Genetics and molecular biology of rhythms in *Drosophila* and other insects. *Adv. Genet.* **48:** 1–280.

CHAPTER 6

Wanderlust

Hence it comes about that flies are not controlled by a sovereign and drillmaster, as are the swarms of bees, those raw recruits. Rather, employing a free-ranging style of campaigning, now they roam about in their foraging.

Leon Battista Alberti (1404–1472)

Fruit flies are not homebodies. They do not make permanent homes; they do not even appear to make temporary homes. Nor are they creatures of habit: They do not repeatedly go back to a favorite haunt in search of food or companionship. They do have preferred habitats and they will make a beeline, as it were, toward an orchard or vineyard to congregate on a pungent, fermenting apple or grape. (Heaven for a fruit fly is vintage time when grapes are gathered and crushed in large vats.) But if given the choice between the familiar and the novel, they will opt for the novel. Unlike Aesop's ant who industriously stores up provisions for hard times, fruit flies are nature's hippies, preferring to move on to someplace new and trusting that they will find food and a place to rest when they arrive. This contrasts with their cousins, the honeybees, who have a stable home (the hive) and go out to search for food in flower patches to bring back to the hive.

Getting Off the Ground

Having been aroused by their internal clock and stimulated by the dawn's early light, the first step in their wanderings is taking off. When a fly decides to take off, it first raises its wings and then jumps by extending its middle pair of legs, while simultaneously depressing its wings (Fig. 6.1). These latter two movements together propel the fly off of a surface.

Once launched, a stereotypical motor pattern takes over to keep the wings beating at a rate of approximately 150–200 beats per second (Hz). Unlike the jellyfish (Chapter 4), whose muscles directly contract the bell to generate force, the fly's mus-

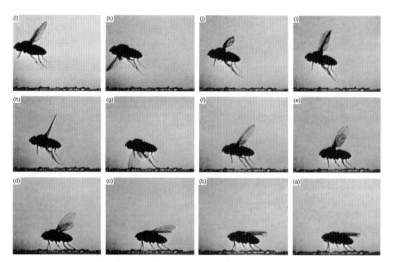

Figure 6.1. Initiation of voluntary flight in *Drosophila* (sequence begins at *lower right*).

cles are not connected directly to its wings. Instead, most insects have evolved an indirect system in which the whole thorax is compressed, and that compression mechanically moves the wings. Compression of the thoracic box occurs as the result of two large sets of muscles working in opposition to each other (Fig. 6.2), the dorsolongitudinal muscles (DLMs) and the dorsoventral muscles (DVMs).

And even more unlike the jellyfish, each DVM or DLM contraction is not driven

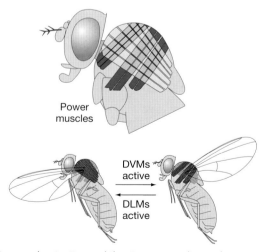

Figure 6.2. Indirect flight muscles in *Drosophila*. Dorsoventral muscles (DVMs) contract on the upstroke and dorsolongitudinal muscles (DLMs) contract on the downstroke.

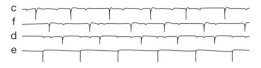

Figure 6.3. Cyclical, sequential firing of DLM motor neurons during flight. Each downward tick is an action potential. Letters (c–e) correspond to muscles labeled in Fig. 6.4.

by an action potential from its respective motor neuron. Instead, these are stretch-activated muscles (like the muscles in our hearts) that are triggered by the initial distortion of the thoracic box. Once activated, the alternating distortion of the thorax by each muscle set is sufficient to activate the opposing muscle set, in an endless cycle until the animal lands. Motor neurons deliver spikes to each of these muscles, but the neurons' primary role is to keep the level of excitation high in general, rather than to trigger each contraction. Nonetheless, these motor neurons are coordinated to fire in a sequential pattern (Fig. 6.3), likely because of reciprocal inhibition (similar to what is seen in leech swimming) (Chapter 4).

If distortion of the thorax is all that it takes to cause the wings to beat, then how does the first distortion occur? This brings us back to the "takeoff" scenario described above, in which the fly first lifts its wings and then simultaneously depresses them as it jumps. The muscle that initiates the first thoracic contraction, the tergotrochanteral muscle (TTM), is under the direct control of a motor neuron (Fig. 6.4). The TTM mimics a DVM and distorts the thorax enough to stretch and activate the DLMs, which cause wing depression (Fig. 6.5). The TTM's motor neuron (the TTMm), in turn, is controlled from the fly's head.

Headless flies do not fly. However, they can actually stand up and right themselves if knocked over. They can perform a grooming reflex if prodded, but they do not fly even if dropped. Flying requires a head. From these observations, we can infer

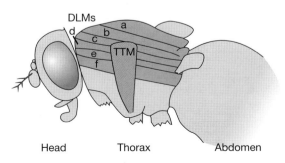

Figure 6.4. Tergotrochanteral muscle (TTM) in relation to DLMs in the thorax.

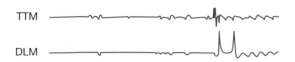

TTM

DLM

Figure 6.5. Sequence of muscle action potentials (TTM, then DLM) preceding initiation of voluntary flight.

that signals originating in the head get communicated to the TTMms. Anatomy suggests that there is an obvious conduit for these signals: a bundle of neurons that includes the largest neurons in the fly, the giant fibers (Fig. 6.6). Electrical recordings show that these neurons do indeed carry impulses from head to thorax in anticipation of takeoff. Simultaneous recordings from a giant fiber as well as from the TTM, DLM, and DVM on the same side show that spikes in the giant fiber precede spikes in the TTM, which in turn precede spikes in the DLM (Fig. 6.5). This sequence of activation is not merely a coincidence, because direct stimulation of the giant fiber produces the same result (Fig. 6.7).

Like the ring giant and motor giant neurons in *Aglantha* (Chapter 4), the large diameter of the giant fibers' axons allows them to conduct action potentials reliably in an emergency, and like the *Aglantha* neurons, they convey signals for routine activity as well as for escape behavior. The same giant fiber/TTM/DLM system also activates an escape response when the fly is startled by a shadow passing overhead, like the barnacle (Chapter 2).

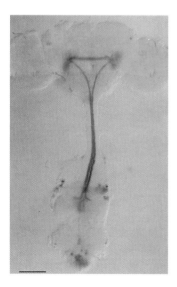

Figure 6.6. The bilateral giant fiber system, largest of a bundle of neurons that project from the brain, through the cervical connective, down into the thoracic ganglion. Scale bar, 100 μm.

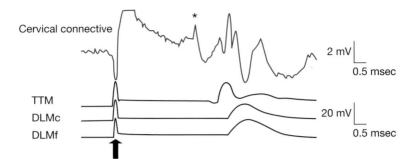

Figure 6.7. Giant fiber (cervical connective) stimulation (*arrow*) produces a giant fiber spike (*asterisk*) that precedes synaptic potentials in the TTM and then in the DLMs. See Fig. 6.4 for anatomical position of the muscles.

Staying Aloft

Once airborne, a fruit fly's major challenge is to stay aloft. Flies have no glide ratio (the forward distance an airplane can coast as it descends toward the ground after losing power). Tipping the scales at roughly 1 mg, they are at the mercy of gusts of wind or even mild air currents that can waft them in unwanted directions or pitch them forward or backward. But fruit flies need to be able to fly steadily and stably.

A sensory organ on the fly's thorax, the haltere, plays the major role in maintaining equilibrium during flight. This small appendage sits on the side of the thorax behind and underneath the wing (Fig. 6.8), where it vibrates rapidly during flight (it has its own little muscles), beating time between wing beats. When the fly makes a rotational movement (e.g., pitching forward or sideways), the rotational force exerted on the halteres causes them to bend, which deforms the fly's exoskeleton. At the haltere's base are tiny sensory structures (campaniform sensilla) that detect these deformations of the exoskeleton. The halteres act as gyroscopes for the fly, sensing rotations and then communicating the direction of movement to the nervous system.

The sensillae have neurons that project back into the thorax, where they contact motor neurons for a set of small, sheet-like muscles lying on either side of the thorax. These are the fly's steering muscles, which regulate the force of contraction exerted by the DLMs and DVMs (Fig. 6.8). The steering muscles modulate the wing beat, so that the fly's flight can be stabilized or change direction.

The steering muscles are not capable of producing much force themselves. Instead, they act like tiny springs, stiffening or relaxing in response to signals from their motor neurons to influence the flexibility of the thorax. By changing the stiffness at different locations in the thorax, the steering muscles subtly alter the amplitude or shape of the wing-beat stroke. The stiffness of a steering muscle, in turn, depends on the timing of its firing, and that timing is what the haltere feedback alters (Fig. 6.9). This specialization is part of the fly's system of indirect flight muscles—the wings are

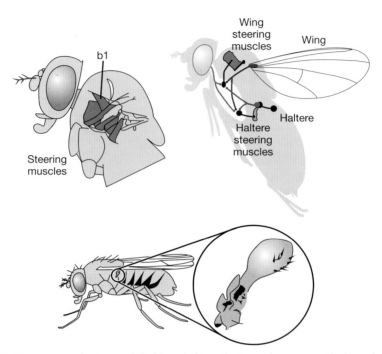

Figure 6.8. Steering muscles (*upper left; blue*), haltere (*bottom*, shown magnified), and schematic of neuronal connections between them (*upper right*). The steering muscle b1 (see text and Fig. 6.9) is indicated (*upper left*).

not directly attached to muscles, and the fly nervous system does not directly activate each contraction of the DLMs and DVMs. As the flight muscles pound out their alternating contractions, the steering muscles nudge them to make subtle shifts in the wing beat; sensory feedback from the halteres induces the nudging. Much of the stability and maneuverability in fly flight is due to this haltere-steering muscle circuit and its ability to compensate. Flies are capable of relatively normal flight even after partial loss of their wings. Damage or lose a haltere, however, and the fly is grounded.

Straighten Up and Fly Right

Whereas the halteres provide a monitoring system for sudden turbulence during flight, vision provides the fly with a means of orienting itself in its visual environment and of correcting for slow deviations or drift during flight. Both kinds of perception allow the fly to monitor self-motion. The fly's visual system takes up most of its head and brain, and one of the system's main functions is to keep the fly stable and oriented during flight.

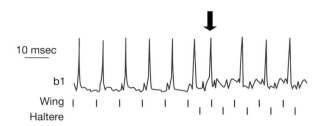

Figure 6.9. Phase advance of a steering muscle following haltere input. (*Top traces*) Action potentials in muscle b1 (see Fig. 6.8). (*Top row of ticks*) Firing of sensory neuron from wing; (*bottom row of ticks*) firing of sensory neuron from haltere. b1 firing follows wing sensory feedback until the haltere neuron begins to fire, at which point (*arrow*) the muscle action potential fires sooner than it would have otherwise. This is a transient effect that soon reverts, and the muscle once again is driven in phase with feedback from the wing. (Recordings were made using the blowfly *Calliphora*.)

The fly's compound eye sees the world very differently from ours, allowing the fly to extract certain kinds of information that humans cannot. They are much more attuned to fast-moving stimuli than we are, and much less attuned to subtleties of spatial pattern. The 700 facets of each compound eye curve around the fly's head so that it can see all but the rear third of its surroundings in the horizontal plane, and more than half of its surroundings in the vertical plane. The anatomy of the visual system allows the fly to take light from the visual field and collect it pixel-like into a two-dimensional array. The first layer of cells in the compound eye is the retina, which functions very similarly to the photoreceptors in the horseshoe crab (see Chapter 2). This layer of photoreceptors synapses onto the first optic lobe, the lamina (Fig. 6.10), where the spatial relationships among points in the visual field are maintained in a two-dimensional set of columns; each bring together the set of photoreceptors aimed at the same point in the visual world. This pixelated, two-dimensional columnar map of the visual world is preserved through a second optic lobe (the medulla) to the innermost optic lobes, the lobula and lobula plate (Fig. 6.10). It is this separation of the visual world into many discrete points, the maintenance of this separation through successive stages of the system, and the ability to compare activity between adjacent points that makes it so sensitive to movement.

Essential to the fly's perception of motion is a set of "wide-field" neurons, cells in the lobula plate that have wide-ranging dendritic trees radiating either horizontally or vertically across the lobe and contacting a full set of columns along their preferred axis (Fig. 6.11). They are poised to sample stimulation of a wide swath of the visual field and track its progression over time. As an additional aid to their detection capabilities, many of these cells are selectively sensitive to movement in one direction.

One set of these neurons, known as "vertical system" (VS) cells, divides the visual field into vertical sectors. Their dendrites radiate in a vertical plane, and when an image moves downward across the eye, the sequential stimulation of a VS cell's den-

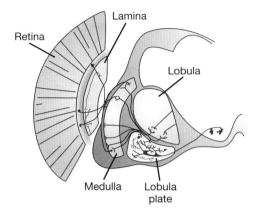

Figure 6.10. Visual system of *Drosophila*, showing representative projection patterns of neurons from the retina through the lamina, medulla, and lobula plate into the brain.

drites (Fig. 6.12) causes it to fire vigorously. An analogous "horizontal system" of HS cells similarly divides the horizontal visual field (Fig. 6.11).

The role of these neurons in flight stabilization was determined from studies in larger flies. Although the anatomy of VS cells would suggest that they mainly respond to vertical movement, in fact they respond best to movements that rotate around an axis near the equator of the eye (Fig. 6.13). How do these cells produce a rotational field response when their dendrites are oriented vertically in the eye? Adjacent cells are coupled to each other in the same way as myoepthelial cells of *Aglantha* are coupled (see

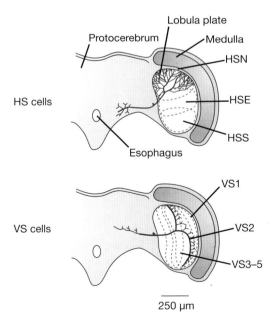

Figure 6.11. Wide-field motion detecting cells of the blowfly lobula plate: (HS) horizontal cells (*top*); (VS) vertical cells (*bottom*). Three HS cells ("north" [N]; "equatorial" [E]; "south" [S]) divide the horizontal field; five VS cells (1–5) divide the vertical field.

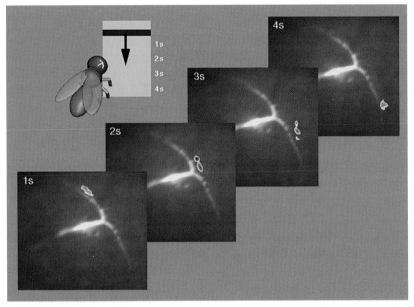

Figure 6.12. Sequential stimulation of dendrites in a motion-detecting cell by a moving bar (*upper left*). The VS2 cell's response to the moving bar is shown by the accumulation of calcium in the cell, made visible by the presence of a calcium-sensitive dye (*yellow* and *red*). The cell is shown in *white*. Each frame corresponds to successive positions of the bar. The anatomical position of VS2 is shown in Fig. 6.11. (Recordings were made using the blowfly *Calliphora*.)

Chapter 3)—through sites on their membrane in which adjacent cells are attached and selectively permeable so that electrical signals can pass from one cell to the next. Connection to the full network broadens the portion of the visual field to which the cell can respond while, in addition, tuning the response to a particular type of motion.

The importance of coupling between VS cells is even more apparent when the fly is shown naturalistic images instead of simple stripes or random dots. When an individual VS cell is cut off from a neighboring cell by, for instance, selectively killing one of them, the remaining cell becomes muddy in its responses and shows no clear axis of rotation. When coupling between VS cells is intact, the cells show stable responses with a clear axis of rotation. Presumably, the patchiness and variable contrast of a naturalistic scene requires a full repertoire of network connections for the fly to adequately detect it. As seen in previous chapters, the action is in the circuit and the network, not in the individual cell.

Returning to the question of how vision might contribute to flight stabilization, VS sensitivity to rotation seems to complement perfectly the halteres' functions, ensuring that the fly will detect deviations from stable flight. When such deviations are too slow to generate a reflex from the halteres, the VS system could kick in. The "horizontal system" (HS) cells (Fig. 6.11) make a different contribution to flight stabi-

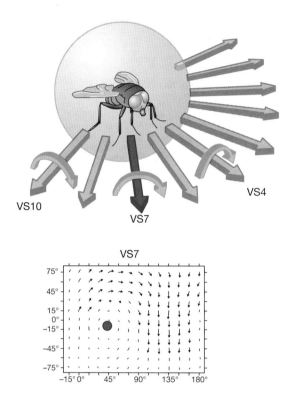

Figure 6.13. VS cells as rotation detectors. (*Top*) A blowfly shown with the various axes of rotation detected by each of its VS cells. (*Bottom*) Receptive field of a VS7 cell. (*Blue dot; lower panel*) Center of the axis of rotation around which this cell is maximally sensitive; (*small arrows*) direction of the cell's maximal response to motion in that part of the visual field. (*Axes*) Position in the visual field as measured by degrees of arc.

lization: the detection of front-to-back movement. Many of the HS cells are sensitive to the direction of movement as well as to its orientation, providing the system with an additional degree of specificity (Fig. 6.14). Finally, many of the HS cells on either side of the brain are connected to one another in an intricate network, permitting the fly to coordinate binocular stimuli. When the HS system is combined with the VS system, it appears that the fly is well endowed with a sensitive and wide-ranging ability to detect self-motion.

How are responses in the optic lobes communicated to the flight-steering muscles? There do not appear to be any direct pathways from the optic lobes to the flight-steering muscles, but there is a direct pathway to the haltere muscles. The wide-field neurons connect to "descending" neurons, such as the giant fiber, that send axons down into the thoracic ganglion. In other words, a moving stimulus to one eye will be conveyed to the haltere muscles and regulate their beating (Fig. 6.8). Changing the rate of haltere beating will then alter the sensory output to the flight-steering muscles and allow for modulation of the wing beat. The end result is that stimuli relevant to flight stabilization, whether the fast mechanical kind or the slow visual kind, are funneled through the halteres and from there to the flight-steering muscles. Clearly, the

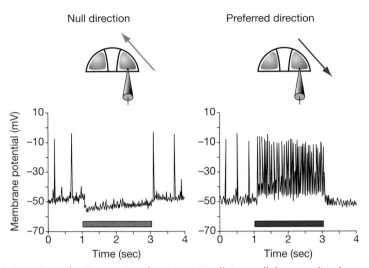

Figure 6.14. Direction-selective, motion-detecting HS cell. Intracellular recording from an HSE cell. The neuron hyperpolarizes and ceases to fire action potentials in response to motion from the back to the front of the animal (null direction). In response to motion along the opposite direction, that is, from the front to the back, the neuron depolarizes and fires trains of action potentials (preferred direction). (*Blue bars; lower graphs*) Time course of visual stimulation. (Recordings were made using the blowfly *Calliphora*.)

network for flight control in the fly runs from one end of its nervous system to the other and then back again. Split-second timing, coordination of action, and distribution of functions over a wide network are all features of the system.

Navigation

Once a fruit fly takes off, where will it go? Because they have a well-developed sense of smell, they are often attracted first by an odor. The smell of fermented fruit is one of their favorites, as you may have noticed the last time you sat outside on a summer evening with a glass of red wine. The smell of fermenting fruit may indicate the location where a male will find females, or vice versa (see Chapter 7). When a fly flies into the air, it may encounter an attractive odor plume. This immediately modifies the fly's flight behavior and its wings beat harder and faster. As it gets closer to the source of the smell, that is, as the smell becomes stronger in the vicinity of the fly, visual cues also come into play. Presumably, the fly is looking for the source of the odor (e.g., a fermenting peach). In a laboratory situation, a fly is unable to locate the source of an odor if it has no visible landmark.

How does the fly track an object? One clue is provided from studies of another

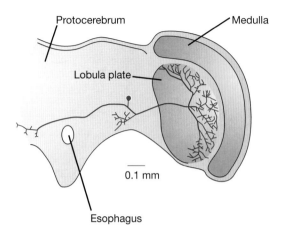

Figure 6.15. Feature-detecting cell (*blue*) in the blowfly lobula plate projecting into the brain.

class of motion-detecting cells in the optic lobes of larger flies. Aside from the wide-field neurons described above, the lobula plate contains another class of neurons known as "small-field" neurons. (They have impressive dendritic trees, but these are less extensive than those of wide-field neurons; Fig. 6.15.) Some of these small-field cells, known as figure-detection (FD) cells, are tuned to detect the movement of small objects relative to their background. The FD cells show little response to motion of the entire field, but they have a vigorous response to a small object moving against its background. These cells have the expected properties for visual tracking, but a direct demonstration of their role in tracking has yet to be made.

Making Decisions

The foregoing description makes the fly's behavior sound very robotic and reflex driven. Although there are strong, reflexive components to flight behavior, the fly does appear to make choices, such as when to take off, what to fly toward, and when and where to land. Motivational aspects of behavior such as these are a challenge for neurobiologists, in part because they require us to guess what may be going on in the animal's brain from the animal's behavior (we have enough trouble figuring out what is going on in each other's brains, even with the help of language). But there are tasks that animals can perform in the laboratory, which, when coupled with analysis of their brain function, give us some indication of what makes them tick. The tendency to anthropomorphize is strong, but this must be restrained by insisting on the demonstration of mechanistic similarities before drawing parallels between us and them.

Some clues have been obtained as to what is going on in the fruit fly's brain when it begins to track toward a visual goal. Tracking behavior can be measured in the laboratory by means of a virtual reality device, in which a tethered fly is present-

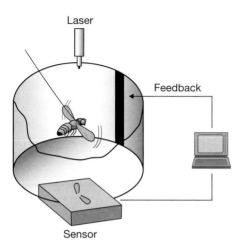

Laser

Feedback

Sensor

Figure 6.16. Fly virtual reality. A fruit fly is suspended (from a wire glued to its back) in the middle of a flight arena, consisting of a cylinder of light-emitting diodes, on which images or patterns are displayed. The fly's wing beats are detected by means of an infrared laser that casts the wings' shadows onto a pair of sensors. The differential amplitude of the left versus the right wing is then fed back through a computer to the image display, allowing the fly to control the position of the image and thus exhibit virtual tracking behavior.

ed with a displayed image (e.g., a vertical stripe), and the differential beating of the left and right wings is measured and fed back to modify the position of the stripe in the arena. This allows the fly to control the position of the image (Fig. 6.16). In this situation, the fly will "track" the image by modulating its wing beats to maintain the image in front of its visual field.

For these studies, flies are genetically engineered so that synaptic transmission can be turned off (and back on again) in different parts of the brain. When this is done in certain brain areas, tracking behavior is selectively impaired. As expected, turning off synaptic transmission from the retina impairs tracking without affecting the ability to fly. Turning off synaptic transmission in a brain center called the "mushroom bodies" also impairs tracking behavior without affecting flight (Fig. 6.17). The fly continues to fly, but it does not keep the object in front of it. The mushroom bodies, so called because of their shape, are located in a brain region implicated in many of the fly's higher cognitive abilities (see Chapter 8). Electrical recordings of local field potentials (LFPs; see Chapter 5) from this region of the fly's brain when it is starting to track show a specific increase in activity when the fly begins to track the stripe (Fig. 6.18). Such activity is present in all experimental flies that are capable of tracking and absent from all of those incapable of tracking.

The decision to begin tracking presumably means that the fly perceives the object

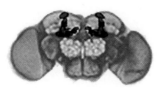

Figure 6.17. Mushroom bodies (*blue*) in the fruit fly brain.

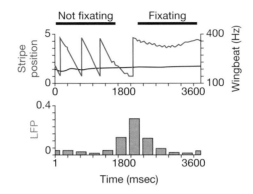

Figure 6.18. Behavioral tracking and brain activity. Shown is the transition to behavioral tracking behavior. (*Upper panel*) Plot of the image position (*blue*) through time, where one ratchet represents a 360° rotation. Tracking is detectable by both the end of the ratcheting and a steadying of the blue trace. (*Black line*) Plot of the wing-beat frequency (Hz), showing that flight is continuous for this whole time period. The transition corresponding to the onset of behavioral tracking is represented by the gap between the lines at the top of the figure. (*Lower panel*) Histogram of the local field potential (LFP) recorded for the same time period as shown for the behavioral data.

and determines that there is something important about it. Registering something important in the environment is one of the major functions of a nervous system and, as seen in *Aplysia*'s response to being attacked by a lobster (Chapter 4), it is performed by neuromodulators. The fly's initiation of tracking is no exception. The neuromodulator in this case is dopamine, and neurons that release it synapse onto the mushroom bodies (Fig. 6.19). Moreover, when these dopamine-containing neurons are genetically engineered so that their synaptic transmission can be turned off, both stripe tracking and the LFP response disappear. In other words, these dopaminergic neurons are as essential to the responses as the mushroom bodies. Presumably, this is because they are all part of the same circuit that registers salience (i.e., something important is in the environment), which is integral to the fly's decision to begin tracking.

If flies are truly novelty seekers, as asserted at the beginning of this chapter, then novelty should play a significant part in their decisions of where to go and what to pursue. By the same token, novel stimuli should be salient to them, as indeed they

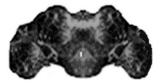

Figure 6.19. Dopaminergic cells (*light green*) that synapse onto the mushroom bodies (see Fig. 6.17) in the fruit fly brain.

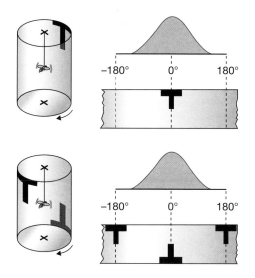

Figure 6.20. Novelty preference. A fruit fly in the flight arena will track a single image (*top*) and, if that same object is presented simultaneously with a novel image, the fly will preferentially track the novel image (*bottom*). (*Left*) Presentation of images; (*right*) relative time spent fixating on the single (*top*) or double (*bottom*) images.

are. Flies in their virtual reality machine will preferentially track toward a novel image if given the choice between it and a familiar one (Fig. 6.20). Moreover, their brains produce the same kind of LFP response to the novel image as when they begin to track toward an object.

Integrating Novelty and Stereotypy

The ability to recognize novelty is one of the cardinal features of animal brains. However, it is one of the more difficult problems for computers to solve: Witness the crudeness with which a credit card company decides to cut off your credit line when it detects "unusual" activity in your account. Aside from perception itself, novelty detection requires that perceived objects be categorized so that similar items are not misperceived as novel. This ability to discriminate also requires memory of previously seen objects so that they may be compared to current ones. And, as shown above for the fruit fly, novelty detection requires registering a novel stimulus as salient.

Stereotypical behaviors, such as the flies' basic flight pattern and its regulation by visual and mechanosensory reflexes, seem to operate in another world altogether from novelty perception and categorization of the surrounding environment. But we can think of the more flexible, cognitive functions as providing a modulating influence on the stereotypical behaviors, sometimes turning them on or off, and other times introducing new elements (e.g., reorientation). The interplay between these two kinds of brain activity and their integration remains one of the most interesting challenges for neurobiologists. Both activities share the property of having evolved because they are adaptive. And by evolving together, they have become tightly integrated.

Further Reading

Borst A. and Haag J. 2002. Neural networks in the cockpit of the fly. *J. Comp. Physiol. A* **188:** 419–437.

Frye M.A. and Dickinson M.H. 2004. Closing the loop between neurobiology and flight behavior in *Drosophila. Curr. Opin. Neurobiol.* **14:** 729–736.

——. 2003. A signature of salience in the *Drosophila* brain. *Nat. Neurosci.* **6:** 544–546.

CHAPTER 7

Love on the Fly

The wren goes to't, and the small gilded fly
Does lecher in my sight.
Let copulation thrive...

from *King Lear* by William Shakespeare

Whereason a fly makes its way to a fallen, rotting peach, or to the rim of a vat of crushed grapes, what is it looking for? Perhaps food, but adult flies are not big eaters. They have already spent their carefree larval youth eating nonstop, day and night. As adults, they may occasionally snack on yeast, but they have more important business to attend to—finding a mate. So when a male fly heads for a rotten peach, he is looking for females. When a female heads for that same peach, she is either looking for males or a suitable place to lay her eggs once having mated. The next time you see a bruised piece of fruit lying on the ground, be aware that you are looking at an active singles scene.

In the world of fruit flies, females generally have the upper hand when it comes to mate choice. Males must prove themselves worthy before a female will consent. Flies are certainly not monogamous in the sense of forming some kind of lasting affiliation between mates, but a female may mate only once or twice in her life, because she will fertilize many eggs with sperm from a single mating. Therefore, the choice matters.

Recognizing a Potential Mate

Fallen pieces of fruit are social centers for many different creatures. Various species of fruit fly may end up together on the same peach, but they will not make the mistake of mating with a fly of the wrong species. Males are not always so successful at making such discriminations at first, and although a male *Drosophila melanogaster* may initially try to court a female *Drosophila virilis* or *Drosophila simulans*, they generally will not get very far. Recognition is crucial and much of it is based on chemical sensing.

Female fruit flies produce a mixture of secreted chemicals on the surface of their external cuticle that acts as a perfume for males (Fig. 7.1). These pheromones are attractive to males and allow them to determine if this other fruit fly is of the same species and the opposite sex. Detection of pheromones takes place on chemoreceptor sensory cells found on the male's forelegs, antennae, and mouthparts (Figs. 7.2 and 7.3). These sensory cells have chemoreceptor proteins and associated ion channels that respond to pheromones. Mutant males that lack these proteins have a hard time courting females in the dark, where chemosensory cues are essential. (In broad daylight, the visual perception of a female arouses a male's interest enough to begin courting.)

The fly's chemoreceptors are of the same family as the opsin proteins in photoreceptor cells (see Chapter 2). They are proteins with seven transmembrane segments, but instead of absorbing quanta of light, they bind pheromones. The binding triggers a conformational change in the protein so that it interacts with a G protein and initiates a cascade of enzymatic reactions. Ultimately, the cascade leads to the opening of ion channels in the membrane of the chemoreceptor cells.

If the male is in the right mood (i.e., if he is old enough but not too old, the humidity is right, and he has not been cooped up with other males—where he would have been exposed to a high dose of male pheromones, some of which inhibit other

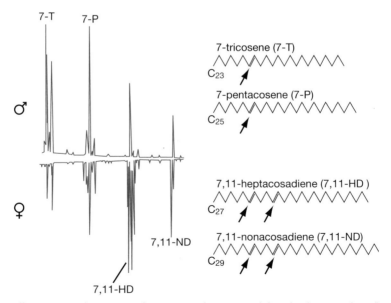

Figure 7.1. Differences in pheromones between male (*top*) and female (*bottom*) fruit flies. (*Left*) Chemical profiles from extracts of male and female cuticular hydrocarbons detected by gas chromatography. (*Right*) Chemical structures of principal peaks from profiles. *Arrows* indicate key double bonds.

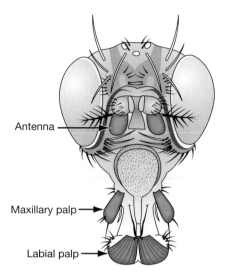

Antenna

Maxillary palp

Labial palp

Figure 7.2. Chemosensory organs on the fly's head: antennae, maxillary palps, and labial palps.

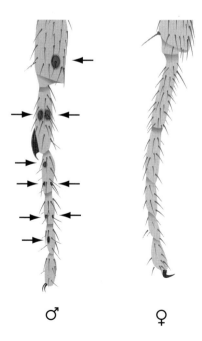

♂   ♀

Figure 7.3. Chemoreceptor (*arrows* pointing to *blue* areas) differences between male (*left*) and female (*right*) forelegs.

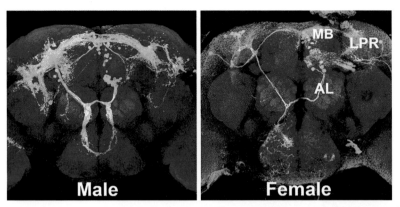

Figure 7.4. Chemosensory circuitry, necessary for male courtship, that differs between males (*left*) and females (*right*). A representative sample of neurons (*light green*) is shown from antennae and palps projecting into the antennal lobes (AL), whose neurons then project to the lateral protocerebrum (LPR) and to the mushroom bodies (MB).

males from courting them), he will sidle up to a female and tap her abdomen with his foreleg. The stimulation of his various chemoreceptors by the female's pheromones will then (usually) trigger his performance of the fruit fly courtship ritual. The signals from the male's chemoreceptors are relayed through the antennal lobes and into the major lobes of the brain, the protocerebrum (Fig. 7.4). When neurons in the lateral protocerebrum are sufficiently excited, the male begins to court vigorously. In males in which these cells have been genetically engineered to be unresponsive, courtship will not begin, and conversely, if these cells are genetically engineered to be more hyperexcitable, the male will start courting at the drop of a hat.

Sex-specific Development in the Nervous System

Some of the circuitry for male courtship is the product of differences in neuronal development in the male brain as compared to the female brain—differences that are both physiological and anatomical. Many of the sex-specific brain regions have been revealed in flies whose brains have been genetically engineered to develop as part male and part female. For example, the fly's antennal lobes are grouped into neuron clusters (glomeruli). Each glomerulus receives a unique combination of synapses from the various types of olfactory receptor neurons. Several of these glomeruli are specifically involved in pheromone recognition. When these glomeruli are engineered to develop as female in an otherwise male brain, the flies fail to distinguish pheromonal differences between males and females.

Surprisingly little of the brain needs to develop in male-like fashion for the triggering of the first steps in courtship: orienting toward the female, tapping her abdomen, and engaging in wing display (in which the male extends one wing laterally and usually follows by vibrating that wing to produce a courtship song). Only one side of the bilaterally symmetrical brain needs to develop as male, and only the dorsal part at that. Sexually mixed flies do not perform fully normal courtship, but they will display these initial steps. (Some investigators have interpreted this finding as confirmation of their suspicion that males use very few neurons when pursuing females.)

Later steps in courtship (Fig. 7.5) require male development in other parts of the nervous system: Production of the song requires male neurons in the thoracic gan-

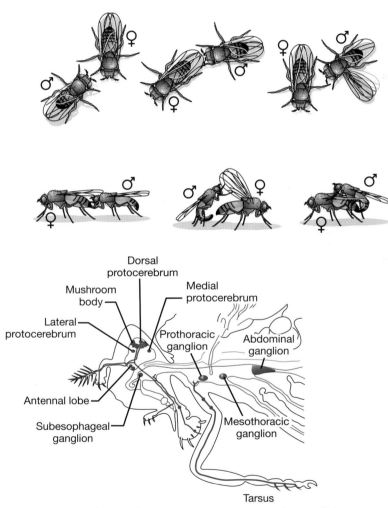

Figure 7.5. (*Top*) Steps in male courtship: orienting, tapping, wing vibration, licking, attempted copulation, and copulation. (*Bottom*) Portions of the nervous system that develop differently in males and females.

glion, licking of the female's genitals requires male brain tissue in the midbrain, attempted copulation requires male neurons in the antennal lobes and the abdominal ganglion, and copulation requires additional male neurons in the abdominal ganglion. Much more of the circuitry for courtship is not sex-specific, however, but, instead, simply makes use of circuits common to many behaviors. These common circuits are present in both sexes and have been incorporated into the courtship routine in males. Courtship is likely to involve a substantial portion of the fly's nervous system.

A gene called *fruitless* is essential for much of this male-specific development. The gene encodes multiple forms of a transcription factor, one of which is male specific. In males that are missing the male-specific product, courtship does not occur. Conversely, when this male-specific product of the *fruitless* gene is expressed in a developing female brain, the adult female will perform male-like courtship behavior that is never otherwise seen in females. The male-specific product of the *fruitless* gene is expressed in most of the brain regions shown to undergo sex-specific development for courtship (see Fig. 7.4).

The Love Song of *D. melanogaster*

One of the major functions of courtship is for the male to advertise himself to a prospective mate. The male does this by extending one wing and vibrating it to produce a courtship song. This song does not sound very musical to our ears (it resembles a monotonous Morse code of rapid bumps), but it is music to the female (Fig. 7.6). When recordings of simulated courtship song are played to potentially receptive females, before placing them in the company of a male, the preexposure makes them more receptive, that is, it takes them less time to begin copulating.

When flies that are part male and part female (see above) contain male neurons in the dorsal brain, they will perform the initial stages of courtship and extend one wing, but they will not produce a proper courtship song unless they also have the necessary male neurons in their thoracic ganglion (see Fig. 7.5). The inference is that they will extend their wing if a signal is sent from the brain down into the thorax (similar to the signal sent to initiate flight; see Chapter 6), but that proper control of the wing's vibration requires fine-tuning from local neurons in the thoracic ganglion, as is also seen in flight control. In fact, one of the flight-steering muscles (b1) beats in time with the courtship song, firing once with every pulse (Fig. 7.7), and may also be the trigger for each pulse.

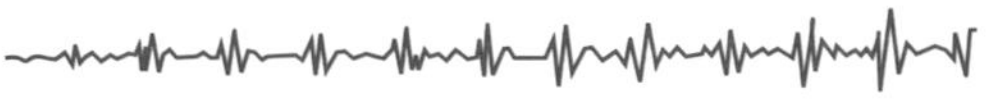

Figure 7.6. Pulses of courtship song produced by male wing vibrations. Each pulse consists of two to three cycles separated by an interval of 32–38 msec.

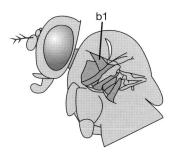

Figure 7.7. The b1 muscle used in the fly's courtship song.

The female "hears" the song through her antennae. The tip of the antenna contains a feather-like structure called the arista that is induced to vibrate by the sound waves from the male courtship song. These vibrations stimulate stretch receptors in the hinge at the base of the antenna, where it attaches to the head, and these stretch receptors convey impulses into the region of her brain required for mechanosensation (Fig. 7.8). The stretch receptor cells contain mechanosensitive channels in their membranes, similar to those in jellyfish (Chapter 3) and microorganisms (Chapter 1).

Flies Have Rhythm

One of the ways that females recognize males of the various species is through the males' courtship song, and a key characteristic of the song that varies between species is its rhythmic timing. The interval between pulses (known as the interpulse

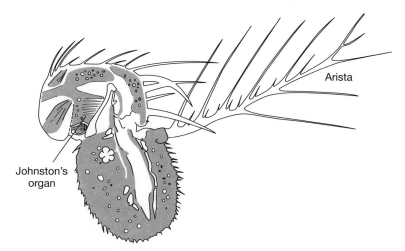

Figure 7.8. Stretch receptors (Johnston's organ) inside the fly's antenna. The arista is on the *right*. The fly's head would be on the *left*.

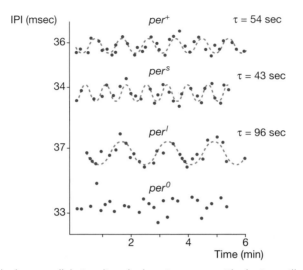

Figure 7.9. Song rhythms parallel circadian rhythms in mutants. Rhythmic oscillation of interpulse interval (IPI) in the fly's courtship song in normal males (*per⁺*) and *period* mutant males: short day (*perˢ*), long day (*perˡ*), and arrhythmic (*per⁰*). τ measures the period of the song rhythm (*per⁰* has no discernible rhythm). (*Dashed lines*) Sinusoidal waves that best fit the data.

interval, or IPI) averages 34 msec in *Drosophila melanogaster*. But closer inspection reveals that this interval actually oscillates in a very regular way. Over the course of approximately 55 seconds, the IPI will gradually increase to 38 msec and then gradually decrease to 32 msec, following a regular, sinusoidal wave (Fig. 7.9).

The physiology underlying this rhythm is not well understood, but it shows a remarkable dependence on the same gene (*period*) that regulates circadian rhythms. Males mutant for alleles of the *period* gene that produce short-day (19-hr), long-day (28-hr), and arrhythmic circadian periods sing courtship songs with correspondingly abnormal rhythms. The short-day flies sing with a 46-second rhythm, the long-day flies sing with a 96-second rhythm, and the arrhythmic flies have no discernible rhythm (Fig. 7.9).

Male flies generally sing their courtship song in bouts of 30 seconds to 1 minute at a time. When a male pauses in his singing, however, he continues to maintain the same rhythm so that when he resumes, the oscillation is right on the beat as if he had been humming it all along. This humming-like behavior requires ongoing firing activity of neurons, as shown by its ability to be interrupted and delayed when action potentials are suppressed (Fig. 7.10). The suppression of action potentials is achieved by means of a temperature-sensitive mutation in the voltage-gated sodium channel of neurons, so that at normal temperature the neurons fire and at higher temperature they become silent. (In the experiment, the fly's entire nervous system was mutant, so it is not known which neurons were specifically responsible for maintaining the rhythm.)

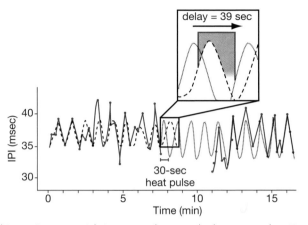

Figure 7.10. Blocking action potentials interrupts the song-rhythm pacemaker. Transient suppression of action potentials was achieved in a male fly whose neurons contain a temperature-sensitive mutation in the voltage-sensitive sodium channel. (*Solid blue line*) Actual singing; (*dashed line [long dashes]*) normal sinusoidal oscillation in the interpulse interval (IPI) up until the moment (*blue dash and inset*) when a 30-sec pulse of high temperature is delivered. During the 30-sec pulse, singing stops and no action potentials can occur. When the male resumes singing, several minutes later, the period of the oscillation is normal (55 sec), but the phase of oscillation of his song is delayed by 39 sec (*shaded box in inset*). (*Dashed line [short dashes]*) Phase-shifted sinusoidal oscillation due to blockage of action potentials. When males not harboring temperature-sensitive sodium channels are similarly treated, the temperature pulse does not delay the phase of their song. When they resume singing, they are "in time," as if they have been keeping the beat all along.

Sequencing in Behavior

Orienting, tapping, wing extension, and singing the courtship song comprise the first steps in a larger sequence of courtship actions. If the female does not run away or kick the male in his face, he will continue to engage in these early steps. Eventually, if she permits it, he will go on to the next steps: licking the female's genitals, mounting her, and copulating. These actions constitute a stereotypical behavioral sequence that is usually performed in order. It is exceedingly rare for a male to skip the early steps and immediately attempt to mount the female. However, when flies are genetically engineered to be mutant for *fruitless* in a central part of the brain called the median bundle, they violate the sequencing rule and skip directly to attempting copulation (Fig. 7.11). The experiment suggests that neurons in the median bundle of males normally inhibit the performance of the late steps in courtship.

Motor Routines and Beyond

Each step in courtship consists of a routine analogous to those described earlier in simpler organisms, such as swimming in the jellyfish (Chapter 3) or flying in the fly

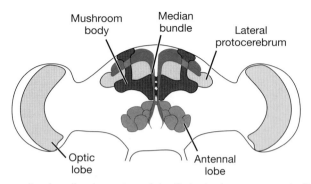

Figure 7.11. The median bundle, the region of the fly brain that causes male flies to skip steps in the courtship sequence when mutant for *fruitless*.

(Chapter 6), but the physiological basis for these steps is far less understood. It is the sequencing of routines such as these that makes behavioral repertoires elaborate and sophisticated. The stereotypical nature of the sequence is a product of natural selection and has been co-selected in the species along with the female's response to it. Understanding the neurobiological basis for the sequencing of behaviors represents a major unanswered question, and its significance may reach far beyond the stereotypical courtship of fruit flies. There is serious speculation that the origin of thinking lies in the sequencing of motor routines. In this perspective, the evolutionary leap to thought came with the ability to string together sequences in the brain such as those controlling motor routines, but without the actual performance. Thus, a behavior as apparently stereotypical as the fruit fly's courtship, with its orderly progression through different actions and its capacity to maintain rhythmic time in the absence of motor output, may have much to tell us about how brains accomplish more innovative and creative tasks, such as thinking things through or telling a story.

Courtship Evolving

In the fly world, choosing the right partner is not a matter of whether the relationship will last; one-night stands are the norm. Instead, mate choice—particularly, picking the right species—means the difference between having offspring or not. As mentioned above, olfactory cues are one means of species discrimination and song is another. Variations in these signals and in the response to them may provide raw material for species divergence. In trying to infer evolutionary mechanisms, one source of clues is to examine the divergence in gene function from one species to the next. Examples of this kind of analysis were presented in the context of geographical differences in temperature compensation or light sensitivity due to the variations in the *period* gene (Chapter 5).

Pheromone Evolution

The fly's chemical signature varies considerably from place to place. Within *D. melanogaster*, the ratios of the different hydrocarbons on the cuticle surface (Fig. 7.1) differ worldwide (Fig. 7.12), and also as locally as between adjacent vineyards near Würzberg, Germany. Differences between species can be even greater, but they do not always reflect phylogenetic distance. In other words, *D. melanogaster* pheromones show greater divergence from those of the closely related species *D. simulans* and *Drosophila erecta* than from those of the more distant species *Drosophila affinis, Drosophila elegans*, and *Drosophila serrata*. Such deviations from expectation are usually indicative of strong selective pressure, suggesting that pheromonal signatures play a major role in speciation.

Fruit flies can be confused when mischievous humans paint them with foreign pheromones. Such treatment is sufficient to induce a *Drosophila mauritiana* male to take the unprecedented step of courting a *D. melangoaster* female, which he would never do otherwise. He will even court a dummy fly, if the dummy is painted with the appropriate pheromones.

Within the *D. melanogaster* species, some of the geographically diverse strains will not court one another. In one case, their reproductive isolation correlates with

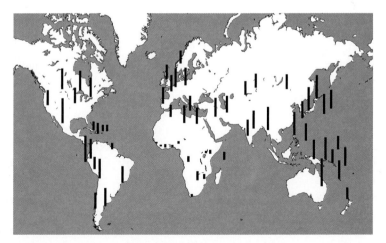

Figure 7.12. Worldwide distribution of levels of the *Drosphila melanogaster* sex pheromone 7,11-heptacosadiene (see Fig. 7.1). (*Black bars*) Relative values for a sample of flies from that locality.

variation in a gene for one of the enzymes involved in pheromone synthesis, *desat2*, that sets the position of the double bond (see Fig. 7.1). A variant form found in Africa and the Caribbean produces much less of the hydrocarbon 7,11-HD that is otherwise predominant in *D. melanogaster* females (Fig. 7.1). The allelic difference between this form of the *desat2* gene and the more distributed form is a deletion of 16 base pairs from the promoter region that prevents transcription of the mRNA for this enzyme.

To test directly for the role of this gene in reproductive isolation, flies were genetically engineered so that the only difference between them was in the *desat2* gene. These two laboratory strains showed the expected difference in hydrocarbon ratios and a mating preference for their own kind. That is, the pheromonal difference was capable of producing a certain degree of reproductive isolation, although not nearly as complete as the original geographical strains' isolation. The incompleteness of effect is actually consistent with genetic mapping studies between the two geographical strains, showing that multiple genes on different chromosomes are collectively responsible for the full-blown isolation between the strains.

Song Evolution

As a song characteristic, the rhythmic oscillation in IPI accounts for a substantial amount of the species specificity of the song. *D. simulans*, as its name implies, is very closely related to *D. melanogaster*, so much so that they can form viable and sometimes fertile offspring. This makes the need for correct recognition all the more pressing (from an evolutionary standpoint). The average IPI of *D. simulans* is approximately 50 msec, in contrast to 35 msec in *D. melanogaster*, and the rhythmic oscillation in *D. simulans* is 35 seconds, in contrast to 55 seconds in *D. melanogaster*. The importance of the two different parameters can be tested by playing synthetic courtship songs to virgin females of either species. These females are then placed with males of the same species and timed for how much more quickly they mate than if no song had been played. When synthetic songs are played to *D. simulans* females, their own species' rhythm is as effective in stimulating them to mate faster with *D. simulans* males as their species' average IPI. Likewise, *D. melanogaster* females care just as much about the rhythm of their own species' song as its average IPI, so both have to be right to enhance their mating. Needless to say, a song that has the correct range of IPIs, but is arrhythmic, is not very effective with these very discriminating females.

Having already seen that mutations of the *period* gene in *D. melanogaster* alter song rhythm as well as circadian rhythm, it stands to reason that variation in the *period* gene between *D. simulans* and *D. melanogaster* might affect the species-specific song rhythm. This turns out to be true. When *D. melanogaster* males are genetically engineered to have the *D. simulans* version of the *period* gene instead of their usual version, they sing a song that has the rhythmic oscillation of *D. simulans*. But this is the only *D. simulans* feature of their song. The average IPI remains unchanged, showing that it is not affected by the *period* gene. Because both species have the same 24-hour circadian rhythm, this too remains unchanged, indicating that selection on the *period* gene has been confined to its song rhythm properties, allowing it to retain its 24-hour timekeeping properties. It may be significant that the segment of the *D. simulans period* gene that confers this song characteristic is approximately the same as the segment affecting temperature compensation among geographically dispersed strains of *D. melanogaster* (see Chapter 5). It is not unusual to find that some portions of a gene are freer to vary than others, presumably because the functions they affect are also freer to vary.

Song is not the only aspect of courtship affected by rhythms. Different species have distinct preferences for the time of day when they are most likely to mate. *D. melanogaster* tends to prefer late in the day and even seems to mate quite well in darkness, whereas *D. pseudoobscura* has two different peak mating times, once around dusk, a few hours later than *D. melanogaster*, and again later in the night (the rogues!). Using a similar approach to the one described above, *D. melanogaster* males were engineered to contain a *period* gene from *D. pseudoobscura*. As you might expect by now, the new strain suddenly developed a predilection for mating rather later than their counterparts who carried an engineered *D. melanogaster* gene (Fig. 7.13).

Is a preference for time of day in mating important for segregating different species? The *D. pseudoobscura*-engineered *D. melanogaster* flies offer a partial test case. They differ only in their *period* genes. If males and females of both types (normal and *D. pseudoobscura*-engineered *D. melanogaster*) are mixed together, will the engineered flies prefer each other, and likewise, will the nonengineered flies prefer each other? So it seems.

Because we already know that the *period* gene also affects courtship song, one might imagine that the observed mating preference was not related exclusively to time of day. By removing the wings from both sets of males, however, any contributing effect of courtship song as influenced by the *period* gene was ruled out. (Wingless males can still court successfully, although it takes them longer.)

What would happen if these two strains were left alone in the laboratory for hundreds of generations? Their temporal separation could lead to the buildup of other differences between them by mutation, and an incipient speciation process could be initiated. Thus, even though they share the same culture bottles, sympatric speciation could occur.

Song variation among species goes beyond merely rhythm and IPI. Some species differ in the notes (i.e., the pulses) themselves. *D. melanogaster* pulses have only two or three peaks that are of relatively small amplitude (Fig. 7.14). *D. virilis* song pulses have more peaks and they are of greater amplitude. A mutant in *D. melanogaster* produces an abnormal song whose pulses resemble those of *D. virilis*. The affected gene is known as *dissonance*, and it encodes a protein involved in RNA processing. (The *dissonance* gene is by no means dedicated exclusively to courtship song. The

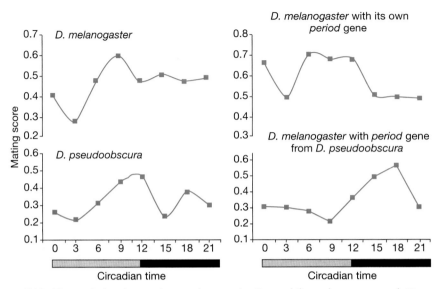

Figure 7.13. Time of day for mating preference in *Drosophila melanogaster* and *Drosophila pseudoobscura* (*left*) and genetically engineered *D. melanogaster* (*right*). Flies were placed in constant darkness so that their circadian rhythms would be governed by their internal molecular clock. Graphs show proportion of pairs mating at various times of day. (*Right*) Strains are genetically engineered *D. melanogaster* with either their same species' normal *period* gene (*upper right*) or the *D. pseudoobscura period* gene (*lower right*). Genetically engineered flies carrying the *D. pseudoobscura period* gene show a peak of mating during the night that is absent in normal *D. melanogaster* and present in normal *D. pseudoobscura*.

protein is expressed all over the nervous system, and other alleles of *dissonance* affect visual sensitivity as well as song. The mechanisms by which it affects either song or vision are not understood.)

Genetic engineering was brought to bear once again and the normal *dissonance* gene from *D. virilis* was substituted into a strain of *D. melanogaster* in place of its own gene. The song in these flies clearly resembles that of the *dissonance* mutant but not that of the normal *D. melanogaster* (Fig. 7.14). Reminiscent of the *period* story, the result suggests that selection for subtle variation in this gene may account at least partially for the species difference in courtship song. The variation must necessarily be subtle so that other functions of the gene, such as those required for visual sensitivity, are left intact.

Neuromodulators and "Instincts"

Behaviors such as fly courtship are sometimes referred to as "hard wired," reflecting the idea that they are innate or instinctual. In fact, male flies raised in total isolation

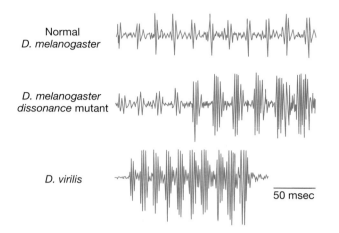

Figure 7.14. Courtship song pulses of normal *Drosophila melanogaster* (*top*), mutants of *D. melanogaster* in the *dissonance* gene (*middle*), and normal *Drosophila virilis* (*bottom*). The *dissonance* mutants produce polycyclic pulses that resemble pulses in *D. virilis*.

from the egg onward are capable of performing the full repertoire of courtship toward a female on their first try. Their behavior may not be quantitatively normal in all respects, but it is qualitatively normal. That does not mean that courtship behavior is entirely impervious to experience (see Chapter 8), but it suggests that, like flying, there has been selection pressure for fruit flies to "know" how to court as part of their normal developmental maturation.

If experience is not essential, then what role, if any, do neuromodulators play in a behavior like courtship? Does a male register the presence of a female with the same internal alarm system (dopamine) that he uses when he begins tracking toward a goal during flight (see Chapter 6)? Apparently it is not the same, because male flies that are depleted of dopamine (after being consistently fed with an inhibitor of its synthesis) will still court females. The salience of a female to a male is thus not dependent on the same neuronal circuitry, a fact that may reflect the species' selection history. The same selection pressures that have produced males capable of courting at first blush have also produced a perceptual response in them that does not need any reinforcement from the neuromodulatory system.

The interplay between "hard-wired" and "plastic" behavior is a recurrent theme in evolution. Most behaviors are a mixture of the two, whereas some are more heavily weighted one way or the other. What is flexible in one species may be stereotypical in another, as seen in the previous discussion of sensitization in *Aplysia* and its relatives (see Chapter 4). In all cases, the behavioral differences reflect variations in the underlying neurons—in their anatomy, physiology, or both. Variation in genes such as *period* or *dissonance* ultimately produce their behavioral effects by acting on one or more of these properties. At present, we still have only a faint picture of very few steps in the mechanisms through which such genes affect the nervous system and alter behavior.

Further Reading

Greenspan R.J. and Ferveur J.-F. 2000. Courtship in *Drosophila. Annu. Rev. Genet.* **34:** 205–232.

Manoli D.S., Meissner G.W., and Baker B.S. 2006. Blueprints for behavior: Genetic specification of neural circuitry for innate behaviors. *Trends Neurosci.* **29:** 444–451.

Peixoto A.A. 2002. Evolutionary behavioral genetics in *Drosophila. Adv. Genet.* **47:** 117–150.

CHAPTER 8

The World As We Find It

If I wrote a book "The world as I found it," I should also have therein to report on my body and say which members obey my will and which do not ...

Ludwig Wittgenstein

How do organisms adapt to the world around them once they find themselves in it? Heredity can get us partially there, but there is no way that the nervous system can be exactly matched to the stimuli and conditions in the world, if only for the fact that those features are always changing. Once out there, it is too late for new heritable gene variations to arise, be selected, and exert an effect. Mutations will only benefit future generations. For the poor fly that has just completed metamorphosis and emerged from the pupa case (the fly's version of a cocoon), time is short. Neural circuits that have already formed during development will limit the range of what the fly can respond to, but accurate matching between the outside world's stimuli and the fly's body and brain requires real-time adjustment.

Traditionally, starting with the theories of Aristotle, insects were assumed to be robot-like creatures that were incapable of learning or modifying their behavior in response to experience. That same assumption also applied to other invertebrates. But beginning in the 1920s with studies of honeybees' ability to remember and communicate the location of pollen sources, and expanding to flies, snails, and others in the ensuing decades (see Chapter 4), it became well established that invertebrate nervous systems register experiences and are altered by them.

Mate Discrimination and Critical Periods

As mentioned before, mate recognition is one of the most evolutionarily important functions that the fly's nervous system must perform. How well tuned the flies are to this task depends on exposure at an early age. Fine-tuning of mate discrimination is particularly important for a South American variety of fruit flies called *Drosophila paulistorum*. These flies exist in what is known as a "superspecies" made up of six "semispecies." That is, they look exactly alike, and they inhabit overlapping territories of the Amazon, Andes, Orinoco, and other central, interior portions of the South American continent, but each semispecies mates preferentially with its own kind.

In the laboratory, *D. paulistorum* is capable of interbreeding, but its tendency to do so depends on exposure during development to the other semispecies. In other words, if flies from the Amazonian and interior semispecies are raised together in a communal situation, they properly discriminate and preferentially mate with their own kind. If, on the other hand, they are raised individually in seclusion, they are much less able to discriminate and exhibit more interbreeding. Exposure of one semispecies to the chemicals produced by the other semispecies during development is sufficient to mimic the effects of being raised communally.

These findings suggest that in order to acquire the ability to discriminate, developing flies must be exposed to the full range of chemicals from both semispecies. As might be expected from the earlier discussion of pheromone recognition (see Chapter 7), the semispecies differ from each other in the composition of pheromone-like chemicals (hydrocarbons) that they produce, chemicals that are likely to be involved in the discrimination process. These flies also have the ability to produce a different cocktail of pheromones, depending on whether they are raised in seclusion or communally.

In the more familiar species *Drosophila melanogaster*, some aspects of courtship behavior also depend on what they are exposed to early on; they are not entirely "hard-wired" (see Chapter 7). Immature males produce some female-like pheromones for the first day or so after they have emerged from metamorphosis. Consequently, these immature males elicit active courtship from mature males, which consists primarily of singing (wing vibration). When these young males are exposed to courtship song as young adults, they become more proficient courters later on when they are mature (Fig. 8.1). As with *D. paulistorum*, the exposure must occur before maturity in order to be effective. In other words, there is a "critical period" for the effect of prior experience.

Critical periods are limited times during which the nervous system is capable of making permanent adjustments to certain stimuli. It can be thought of as developmentally regulated learning, confined to a particular stage, and it contrasts with the garden variety learning that can occur at any time (such as in *Aplysia*; see Chapter 4). In addition to a critical period, courtship also has aspects of the everyday kind of learning, despite the fact that it is generally stereotypical (Chapter 7). The most prominent of these aspects is the behavior of a male fly in relation to a female who has already mated.

During copulation, before the transfer of sperm, the male transfers a mixture of peptides to the female in the seminal fluid. These peptides eventually alter the female's behavior, so that several hours after copulation she becomes unreceptive to subsequent advances from males and begins active egg-laying. Her synthesis of pheromones also changes. The upshot is that after 10–20 minutes of courting one of

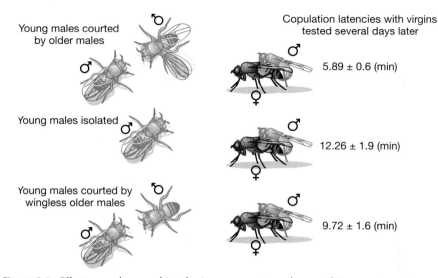

Figure 8.1. Effect on male courtship of prior exposure to male courtship song. Immature males are placed with mature males (*top*), left alone (*middle*), or placed with wingless older males (*bottom*) who produce a faint courtship song vibration that is less effective than a full courtship song produced by males with intact wings. When these treated males mature, they are placed with receptive, virgin females and the average time required to achieve copulation (*column on right*) is measured. Prior exposure to normal song enhances the treated males' success in copulation.

these unreceptive females, the male abruptly stops. If he now runs into another female, even if she is potentially receptive, he is less likely to try courting her. This syndrome (all too familiar to males of many species) is known as courtship depression, and it lasts for up to 2 hours.

The adaptive value of courtship depression is not immediately obvious. Why should a male give up any possibility of finding a mate? One possible explanation is that courtship (especially the final stages in which the male attempts to mount the female) is energetically very expensive for the male. If he is to appear vigorous and suitable for a female, he cannot waste his energy on unproductive ventures. This could be particularly important when flies are living in relatively high density and the probability of running into a mated female is relatively high.

The mated female emits inhibitory signals that the male perceives through sensory cells on two small appendages called the maxillary palps that can be found on either side of his proboscis (Fig. 8.2). In mutant males lacking sensory cells on the palps, or in males whose palps have been surgically removed, courtship of mated females is uninhibited. The sensory neurons from the palps send their axons to the same part of the brain, the antennal lobes, that in turn receive input from the antennal sensory neurons. Neurons from the antennal lobes then project to the brain centers regulating courtship (see Chapter 7).

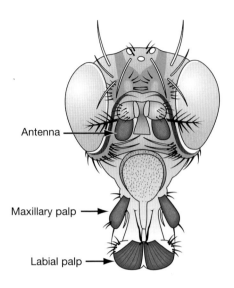

Antenna

Maxillary palp

Labial palp

Figure 8.2. Chemosensory organs: antennae, maxillary palps, and labial palps.

Depression of the male's courtship behavior and the persistence of its depression depend on mechanisms of synaptic modification similar to those used in *Aplysia* sensitization (see Chapter 4). Males mutant in the nervous system for adenyl cyclase, a form of the enzyme that makes cAMP (cyclic AMP) (Fig. 8.3), do not stop courting a mated female at all and will readily court subsequent females they encounter. Mutants in cAMP phosphodiesterase, the enzyme that degrades cAMP and thus regulates its levels, show the normal cessation of courtship of the mated female, but then "forget" immediately and will try to court the next female they see. As described previously (see Chapter 4), cAMP acts through protein kinase A, which modifies proteins in the cell such as potassium channels and transcription factors. Another protein kinase involved in synaptic modification is also required for courtship depression: Ca^{2+}/calmodulin-dependent protein kinase II (CaM kinase). Males genetically engineered to have reduced levels of CaM kinase fail to be inhibited by mated females and will subsequently court other females.

The "early" and "late" stages of courtship depression require synaptic modification in different parts of the brain. When males are genetically engineered so that the CaM kinase deficit is restricted to particular structures in the brain, the two separate effects on courtship—initial depression (early) and persistence of the depression (late)—can be affected separately. When parts of the brain involved in detecting odors (the antennal lobes or lateral protocerebrum) are targeted, initial depression and subsequent persistence are both affected (Fig. 8.4). However, when parts of the brain involved in decision-making or motor control (the mushroom bodies or central complex) are targeted, both of which lie anatomically downstream from the first two

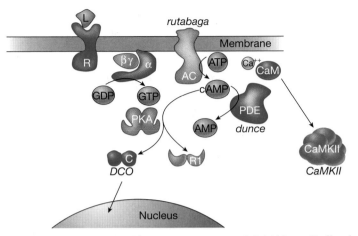

Figure 8.3. Genetic variants in cAMP pathway components and CaM kinase II affect learning. The enzyme adenyl cyclase (AC) is activated by the G-protein α subunit, causing it to dissociate from its β and γ subunits, after its associated receptor (R) binds its neurotransmitter ligand (L). The enzyme's activity also depends on binding calmodulin (CaM) complexed with Ca^{2+}. AC then synthesizes cyclic AMP (cAMP), which binds to protein kinase A (PKA), causing its regulatory subunit (R1) to dissociate, and activating its catalytic subunit (DCO). The enzyme cAMP phosphodiesterase (PDE) degrades cAMP.   The *rutabaga* mutant is in the gene for adenyl cyclase, the *dunce* mutant is the gene for cAMP phosphodiesterase, and the *DCO* mutant is in the catalytic subunit of PKA, all of which are defective in courtship conditioning. Genetic variants with reduced calcium/calmodulin-dependent protein kinase (CaMKII) activity are also defective in courtship conditioning.

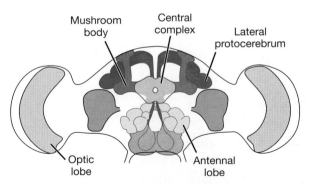

Figure 8.4. Brain areas showing modification for courtship depression: antennal lobes, lateral protocerebrum, mushroom bodies, and central complex.

regions, initial depression occurs but the effect does not last. It is as if the olfactory structures are necessary for training, whereas the other ones are required for consolidation of the training into a retrievable memory.

Odor Learning in the Honeybee

In Chapter 6, flies were described as novelty seekers, preferring to see what is out there rather than keeping to old familiar haunts. Such behavior contrasts with honeybees who must return to the same home hive regularly and also be able to locate and return to flower patches that are rich sources of pollen. They exploit a single floral species as long as it provides a profitable nectar and pollen reward. Therefore, they must recognize and find flowers that they have already identified during previous foraging runs.

Consistent with this difference between the two insect species, it is much easier to get a honeybee to remember something for a long time than it is a fruit fly. In the laboratory, if an immobilized honeybee is presented with an odor and promptly given some sugar as a reward (Fig. 8.5), and if this pattern is repeated a few times, the bee will henceforth respond to that odor by extending its proboscis (mouthparts) in anticipation of more sugar (Fig. 8.6), and the effect will last for days. (In Pavlov's terms [see Chapter 4], the odor is the conditioned stimulus [CS], the sugar is the unconditioned stimulus [US], and the extension of the proboscis after sugar presentation is the unconditioned response [UR]. After training, the odor-induced proboscis extension is the conditioned response [CR]. The order of events makes a big difference to the bee; the odor must precede the sugar or no association will be made. In other words, the CS, that is, the odor, becomes salient when it takes on predictive power.)

It is easy to imagine the adaptive value of this response. If a forager honeybee discovers an attractive flower patch, she will land on a flower to check it out. Any nectar

Figure 8.5. Honeybee extending its proboscis to drink sugar water.

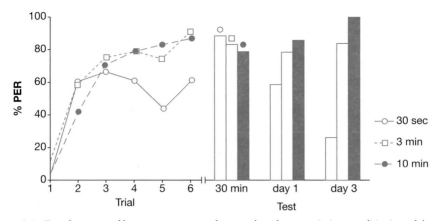

Figure 8.6. Development of long-term memory for an odor after associative conditioning of the proboscis extension response (see Fig. 8.5). (*Left*) Increase in proboscis extension response (PER) with successive training trials; (*right*) persistence of memory after training for 30 min, 1 day, or 3 days. Intervals between training trials were 30 sec (*open circles*), 3 min (*open squares*), or 10 min (*filled circles*).

present on the flower acts as a positive reinforcer for the value of that flower's odor, and so the bee associates the two. If, in the vicinity, the bee then encounters additional flowers with nectar of the same odor, then the reinforcement is stronger and lasts longer. In the laboratory, this process is mimicked by delivering the odor and sugar at intervals of a few minutes (Fig. 8.6). If the intervals are too short (30 sec), memory decays after 2 days. If intervals are a few minutes, memory lasts for at least 3 days.

The dependence of memory on the intervals between odor and reward experiences may be related to the natural foraging sequence of the honeybee. When a bee visits the individual florets of a compound flower in rapid succession, and the intervals between experiences are only a few seconds, memory will be transient. But if the bee leaves the flower patch to go back to the hive and then returns to the patch after a few minutes, memory will have improved, because of the elapsed time between visits. If the flowers last for days in the field, the bee's memory will improve even more.

The circuitry for the odor-sugar proboscis extension response uses olfactory sensory neurons on the antenna that project into the antennal lobes, whose neurons then project to the lateral protocerebrum and mushroom bodies (Fig. 8.7). As in the fruit fly, these are all sites of plasticity. The mushroom bodies appear to be particularly important for sorting out the discriminations, as opposed to receiving primary sensory stimuli. The sugar response enters through the gustatory (taste) cells in the mouthparts, antennae, and legs, and stimulates a widely branching neuron called the VUM (ventral unpaired median) neuron, which branches throughout the brain, contacting all of the structures involved in olfactory response. This cell's firing correlates with the presentation of sugar and also with extension of the proboscis.

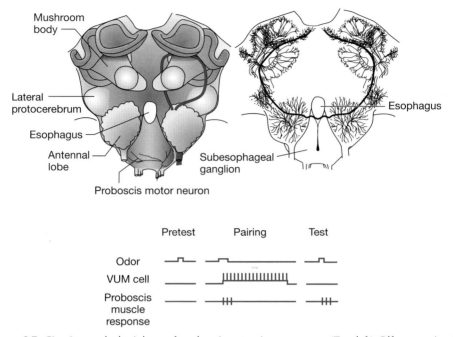

Figure 8.7. Circuitry and physiology of proboscis extension response. (*Top left*) Olfactory stimuli enter through antennal lobes and travel to lateral protocerebrum and mushroom bodies. The proboscis motor neuron drives the proboscis muscle to extend the mouthparts. (*Top right*) Reinforcement stimuli enter through proboscis to subesophageal ganglion to the VUM cell, whose wide-ranging projections are shown extending all through the antennal lobes, lateral protocerebrum, and mushroom bodies. (*Bottom*) Direct stimulation of the VUM cell during presentation of an odor substitutes for the unconditioned response and produces a bona fide conditioned response. Pretest, initial presentation of odor (conditioned stimulus); pairing, simultaneous presentation of odor and stimulation of VUM cell (unconditioned stimulus); test, presentation of odor now elicits proboscis extension by activating proboscis muscle (conditioned response).

The VUM neuron appears to be the actual mediator of the reward (US). Direct stimulation of the cell can actually substitute for sugar (US) presentation (Fig. 8.7, bottom). If the VUM is stimulated at the same time as odor is presented, or just after, it is as effective in training as sugar presentation, and the bee will learn the odor as if it were associated with a sugar reward. VUM cells release the neuromodulator octopamine (Fig. 8.8), which appears to function analogously to serotonin in *Aplysia* (see Chapter 4). Injection of octopamine into the appropriate sites in the bee's brain can also substitute for the sugar reward, provided that the injections occur coincident with odor presentation. These findings strongly implicate the VUM neuron in reward.

An example of how the honeybee brain changes as a consequence of odor-reward training can be seen in the pattern of responses in the antennal lobes. These lobes are divided into clusters called glomeruli, which receive synapses from a

Figure 8.8. Octopamine.

unique combination of olfactory receptor neurons, as in *Drosophila* (see Chapter 7). Different odors excite different groups of glomeruli, and training causes glomeruli to become more excited by the odor they prefer (Fig. 8.9). Eventually, additional glomeruli are excited as well, altering the pattern as well as the intensity of the olfactory signal.

The Seasons of Memory

Fruit flies are also capable of odor learning in the laboratory, though not with the same élan as honeybees. Effective learning in fruit flies requires the stick rather than the carrot, so instead of a nice sugar water reward, the poor little flies are given an electric shock instead. But they tolerate it and come out the wiser. Odor learning in both fruit flies and honeybees uses the familiar cAMP/protein kinase A cascade in their neurons to produce the same kind of synaptic and behavioral changes as seen in the modification of fly courtship (see Fig. 8.3 and Chapter 4).

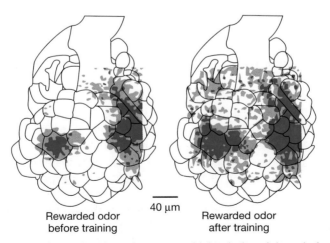

40 µm

Rewarded odor
before training

Rewarded odor
after training

Figure 8.9. Activity of glomeruli in honeybee antennal lobes before (*left*) and after (*right*) training. Images are of calcium levels in glomeruli in which *light blue* reflects moderate and *dark blue* reflects high levels, which correspond to moderate and high levels of ongoing neural activity.

Whether these insects will retain the effects of their odor training for hours or for days depends on their training regimen. Like students at exam time, if the training is all crammed into one long session, they remember it well for a short time but then lose it. If, instead, they train at regular intervals, the effect lasts much longer. The molecular result of interval training is to activate protein kinase A for a long period of time, which ultimately allows it to activate the CREB (cAMP responsive element binding) protein, a transcription factor that enters the nucleus and alters gene expression. Flies that are genetically engineered to produce an inhibitory form of CREB (one that blocks gene activation in the nucleus) are incapable of remembering for more than a few hours even if trained properly. Thus, it is the synthesis of new proteins after CREB activation that modifies the cell for a long-lasting effect. The selfsame mechanisms lie at the heart of the synaptic plasticity responsible for long-term changes in our own brains.

What are these new gene products that are so effective in producing long-term changes in neurons? One group comes from a class of mechanisms that regulate the localization and translation of mRNAs (messenger RNAs) in the cell. These proteins are responsible for sequestering specific mRNAs into particular compartments (such as the fine branches, i.e., dendrites, that receive synaptic contacts from adjacent cells). Several of these proteins are newly made as the result of interval training. Flies that produce a mutant version of one of these proteins, called Staufen, show a failure to remember for long periods of time after interval training. The effect is similar to CREB inhibition. The inference of these findings is that local production of certain proteins helps to stabilize changes at particular, stimulated synapses.

The Varieties of Cognitive Experience

Honeybees are clearly very effective at learning and remembering odors but they are capable of much more. Their greater sophistication is revealed when they can fly freely, examine different visual patterns, and choose which pattern to approach. In this situation, they can be trained to associate a particular visual pattern with a reward (sugar water, as usual). Bees have no trouble discriminating different specific patterns. More impressively, they can learn to distinguish categories of pattern, such as symmetry versus asymmetry. In one such test, freely flying bees were presented with three different patterns on the ground in an enclosed area. One pattern was symmetrical and the other two asymmetrical, and a sugar water reward was placed only at the symmetrical site (Fig. 8.10). They were then presented with another triad of patterns, one symmetrical and two asymmetrical, and again rewarded at the symmetrical site. After eight groups of these stimuli were presented for dozens of times, the bees were presented with a novel group of patterns, some symmetrical and some asymmetrical, and observed as to which one they approached. If trained to prefer symmetry, they approached almost exclusively the symmetrical test patterns, even though these patterns were novel for the bees. If trained to prefer asymmetry, they

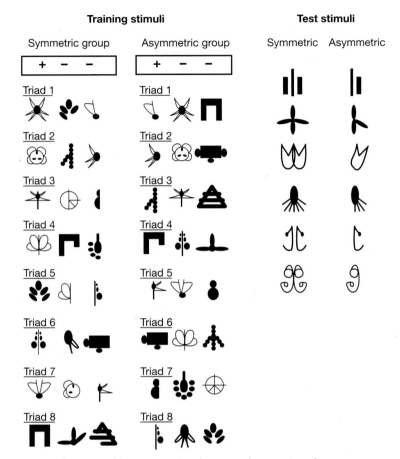

Figure 8.10. Honeybees are able to recognize the general categories of symmetry vs. asymmetry, independently of the stimuli used. (*Left*) Individual honeybees were trained on eight successive triads of stimuli, either for symmetry or asymmetry. Each triad of the symmetric group consisted of a rewarded (+) symmetric stimulus and two different unrewarded (–) asymmetric stimuli presented at the same time. Each triad of the asymmetric group consisted of a rewarded (+) asymmetric stimulus and two different unrewarded (–) symmetric stimuli. (*Right*) During tests, novel stimuli (symmetric and asymmetric) were presented. None was rewarded. By the end of the training, bees were making correct choices more than 80% of the time.

showed a preference for asymmetry. This result suggests that honeybees are capable of a kind of abstract categorization.

An even more abstract category that bees can learn to recognize is "difference" versus "sameness." In these tests, bees were initially presented with one stimulus (e.g., blue) followed by a choice between two test samples, one of which was the same (blue) and the other different (gray; Fig. 8.11). In a subsequent test, the stimuli were changed, but the principal (e.g., sameness) was retained. Eventually, they learn

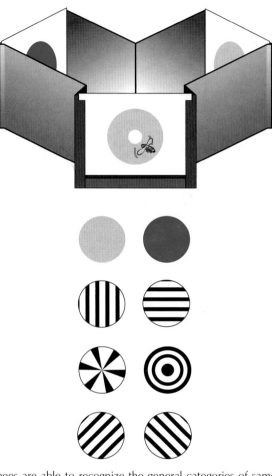

Figure 8.11. Honeybees are able to recognize the general categories of sameness and difference, independently of the stimuli used. Bees were presented with a sample and then with a set of stimuli, one of which was identical to the sample. The bees had to learn to always choose the stimulus that was identical to the sample, despite the fact that the sample was regularly changed. A Y maze was used for training and testing the bees. The sample was presented at the entrance of the maze and bees passed through without being rewarded. The honeybee was rewarded with sucrose solution if it chose the stimulus presented on the back walls of the maze that was identical to the sample. The *gray* color of the first stimulus was yellow in the actual experiment.

to consistently approach the test sample that was the same as the initial sample, even if all were novel. To test the generalizability of this response, they were then tested on line patterns (Fig. 8.11) in the same way: One line pattern was presented first, followed by a choice between the same pattern and a different one. Bees trained to recognize sameness in color prefer sameness in line pattern; those trained to recognize difference in color then prefer difference in line pattern.

If you are not yet convinced of the honeybee's intelligence, consider their ability to learn a map of an area to the point that they can find novel, shortest routes between two points that they have not previously navigated or seen in one panoramic view. When trained first to go to one distant feeder (S_a) from their hive, then to another (S_m), and then placed in a novel site between them (S_3), the bees found their way back to the hive by a direct route, even if they could not see the entire terrain (Fig. 8.12).

Even more impressively, when bees are prevented from following a familiar route between the hive and a food source, they display a rich knowledge of the surround-

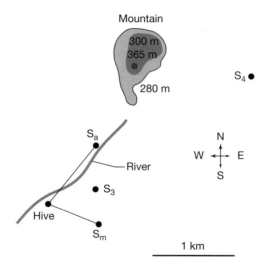

Figure 8.12. Navigating novel routes by honeybees. A group of bees was trained to collect a sugar solution at location S_m in the morning and at location S_a in the afternoon, 0.6 km and 0.8 km, respectively, from their hive. A prominent distant landmark was a cone-shaped mountain standing 150 m above the wide and flat agricultural area surrounding it. S_3 and S_4 were the novel release sites. S_3 was halfway between S_m and S_a, whereas S_4 was far away (3 km) from both. At the new release site S_3, halfway between S_m and S_a, bees departed toward the hive, choosing a novel flight path that they had never flown before. However, bees displaced too far, to S_4 (a more distant release site), were unable to find a new correct route, probably because at the new unrecognizable release site, S_4, the memories associated with S_a and S_m could not be activated simultaneously, and the bees were therefore unable to interpolate a new flight path.

ing space. When they are aiming at a particular goal, displacing them to a novel site will not impede their ability to reach that goal, provided that the new site is in the same general area. Not having a known route to follow, they use their memory of the whole area as if it were a global map. It is as if they are able to attach a coordinate location to each site in space, thus enabling them to develop novel routes. In other words, they follow a previously unknown route between two locations based on knowledge of the relationship of each site to its global space.

This kind of place learning represents a degree of sophistication in cognitive ability that would impress even Aristotle. In the jargon of experimental psychology, map learning is a form of episodic memory, a kind of memory considered to be at the pinnacle of the cognitive scale, exceeded only by our human propensity for language. Such a mapping ability complements the honeybees' aptitude, already described, for categorization according to abstract principles. Add to this their possession of a sophisticated communication system, the "waggle dance" by which foragers communicate the location and quality of pollen sources to their compatriots in the hive, all occurring in a brain with a mere 1 million neurons, and you have an organism that merits our respect.

Probing the Physiology of Cognition

The mapping abilities of honeybees were definitively proven using transponders—tiny radio transmitters attached to the bees that allowed their entire flight path to be accurately traced. As a technical feat, that is impressive enough. The possibility of recording from their brains while they are navigating in order to understand the physiology underlying their cognitive prowess, however, strains the imagination, at least for now. A more accessible (if less charming) experimental organism is the cockroach, an insect that shows some ability to learn locations and from which brain recordings can be made while it is walking around.

Although not as impressive as the honeybee, cockroaches show some degree of place learning when subjected to a hidden target test. Cockroaches were placed on a plate that could be heated to a temperature that they find unpleasant. An unmarked sector of the plate remained unheated, and the only way to learn its position was by referring to visual landmarks on the walls of the chamber. Cockroaches could learn the position of the privileged sector and find it again more quickly using landmarks (Fig. 8.13). Their ability to find a hidden target using landmarks is blocked by surgical lesions to their mushroom bodies. These lesions, however, do not debilitate them in general, because they can still readily find a visible target, even if they cannot remember it.

When one looks into the cockroach's mushroom bodies to see what its neurons are doing, one finds versatile cells—some that respond to multimodal (visual and tactile) sensory input (Fig. 8.14) and others in which activity is predictive of a particular kind or direction of movement (Fig. 8.15). A multimodal response indicates that the

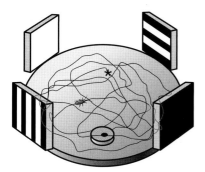

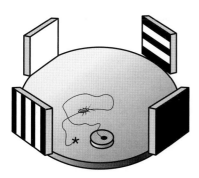

Figure 8.13. Place learning in the cockroach. The insect was placed in a heated 30-cm-diameter arena with landmarks at each corner. An invisible target (*the internal blue circle*) was kept at low temperature. Tracks represent the route taken by the cockroach before finding the privileged sector early (*top*) and late (*bottom*) in its training.

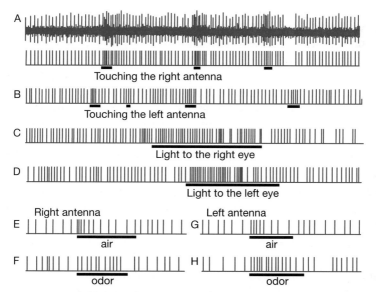

A

Touching the right antenna

B

Touching the left antenna

C

Light to the right eye

D

Light to the left eye

Right antenna

E

air

Left antenna

G

air

F

odor

H

odor

Figure 8.14. Action potentials recorded from a multimodal mushroom body neuron in the cockroach brain. Spike frequency increases after stimulating antennae on either side by touch (*A,B*), a light stimulus to either eye (*C,D*), air (*E,G*), or odor (*F,H*).

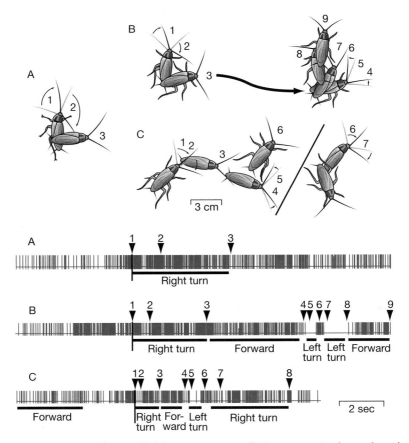

Figure 8.15. Action potentials recorded from a motor-predictive neuron in the cockroach mush-room body. (*Top*) Movements made by a walking cockroach (numbers represent sequential positions in each episode); (*bottom*) action potentials recorded in the motor-predictive neuron correspond-ing to steps in turning sequences. *A*, *B*, and *C* spikes (*bottom*) correspond to *A*, *B*, and *C* movements (*top*). The neuron's activity anticipates the movement and continues through to its end.

cell is integrating many kinds of signals. Although it is not known whether these par-ticular cells are critical to the cockroach's place learning, they have the characteris-tics that one would expect. The other class of mushroom body neurons, with specif-ic motor-predictive properties, is characteristic of the kinds of cells known to be involved in spatial learning in rats and mice. Anatomically, the mushroom bodies in all insects occupy a central position in receiving multiple kinds of sensory inputs and sending out signals to motor control areas. Although suggestive and intriguing, these features of neurons only serve as loose suggestions for a vague outline of how processes such as place learning and spatial memory might work. A complex and wide-ranging network of cells must be involved.

Memory in general presents a major challenge to our understanding. Although it is clear that memories require synaptic modification in certain parts of the brain, as seen above in both *Aplysia* and *Drosophila*, is that process adequate to explain the process of remembering? What exactly constitutes a memory physiologically, and how do the interactions among various brain regions make possible the consolidation and retrieval of a memory? These remain open questions. The word "retrieval" itself may be misleading because it implies that a stable, unvarying entity is being obtained from somewhere, as if taking a book off a shelf. The reality of memory is that it is a highly variable and sometimes unreliable process. Its capriciousness has been well documented in humans, rats, crabs, and sea slugs. We have all experienced it. It seems likely that imperfect memory may be true of all creatures that have memories.

Further Reading

Hammer M. and Menzel R. 1995. Learning and memory in the honeybee. *J. Neurosci.* **15:** 1617–1630.

Mehren J.E., Ejima A., and Griffith L.C. 2004. Unconventional sex: Fresh approaches to courtship learning. *Curr. Opin. Neurobiol.* **14:** 745–750.

Menzel R. and Giurfa M. 2006. Dimensions of cognition in an insect, the honeybee. *Behav. Cogn. Neurosci. Rev.* **5:** 24–40.

Mizunami M., Okada R., Li Y., and Strausfeld N.J. 1998. Mushroom bodies of the cockroach: Activity and identities of neurons recorded in freely moving animals. *J. Comp. Neurol.* **402:** 501–519.

Schwarzel M. and Muller U. 2006. Dynamic memory networks: Dissecting molecular mechanisms underlying associative memory in the temporal domain. *Cell Mol. Life Sci.* **63:** 989–998.

CHAPTER 9

Are All Brains Alike? Are All Brains Different?

Man is the only animal that blushes. Or needs to.—*Pudd'nhead Wilson's New Calendar*

Mark Twain

Brains make it possible for all of us, human and invertebrate alike, to move through the world and deal with its kaleidoscope of changing conditions. The world is actually a bewildering place and our senses are constantly bombarded with a chaotic avalanche of electromagnetic and tactile signals. The task of making "sense" of all of this is left to our brains and they do such a good job that we do not even notice it. In contrast, people who have been blind or deaf from birth and who then acquire the missing sense (through surgery or prosthetic devices) report that initially it is all chaos and only after some time does it begin to "make sense." There is no reason to think that the situation is altogether different for our invertebrate friends. The main distinctions would be due to differences in the arrangements of their bodies and the resulting differences in what they receive through their senses, as well as their more limited neural repertoire for "making sense."

The foregoing examples give a taste of how some of the better studied invertebrates do it. Each of the mechanisms outlined has general relevance beyond both the particular task and the particular species. Often, the molecules performing the mechanisms are similar, as are their arrangement and their interactions with one another. Other times, omissions, additions, or substitutions modify the mechanism from one species to another. The products of evolution are variations on many themes.

For all the molecular and mechanistic similarity that we find at the level of the neuron, there is an equally striking divergence at the level of anatomy. The nervous system of a jellyfish is laid out in a pattern that bears little resemblance to that of *Aplysia*, and neither of them looks at all like a fruit fly brain. (Their body plans do not have much in common either.) When invertebrates are compared to vertebrates, the same similarities at the molecular and cellular level are found, but even greater anatomical disparities are evident. Vertebrate brains have almost no anatomical homology to those of invertebrates. In contrast, all vertebrate brains have a high degree of homology to one another, and there is a fair degree of homology within invertebrate phyla, e.g., all arthropod brains bear a certain resemblance to one another. Anatomical homology generally means considerable conservation of neural circuitry.

Despite the anatomical differences between invertebrates and vertebrates, there are some uncanny behavioral resemblances. These include major similarities in the phenomenology of their circadian rhythms, sleep, associative conditioning, and memory phases, and hints of similar aspects of motor control, arousal, and perceptual discrimination and categorization.

How can such divergent anatomies give rise to such similar behaviors? Part of the answer surely comes from the homology of cellular mechanisms already mentioned. This is particularly important in circadian rhythms, where the timekeeping mechanism occurs inside individual cells that then act as pacemakers for the brain. It is equally relevant to the plasticity necessary for learning, where the synaptic mechanisms are localized to individual cells. But behaviors such as motor control, place learning, discrimination, and categorization are higher-level mechanisms that require complex circuitry. In these cases, any behavioral similarity requires a different explanation.

A Common Strategy?

Is each type of nervous system a thing unto itself? Do the variations between species represent different arrangements and functional strategies? Or do the differences camouflage an underlying commonality?

Perhaps all nervous systems make use of common general strategies. Anatomical disparities may mask underlying functional similarities in the tasks performed by various circuits. One obvious case of functional similarity, despite anatomical divergence, is in the role of neuromodulators. In both vertebrates and invertebrates, many of the cells releasing transmitters such as serotonin and dopamine are diffusely distributed in the brain and do not synapse closely onto other neurons. Release of the modulators is thought to occur, at least in part, in a diffuse, nonlocal manner, like a sprinkler system.

These modulators influence reward, aversion, arousal, depression, and other general states that affect behavior of the animal qualitatively. For this reason, the cells and circuits mediating these effects are sometimes referred to as "value" systems. The combination of modulating other neurons' activities, being diffusely distributed, and affecting qualitative states gives these systems a unique role in the nervous system. Their nonspecific action, combined with the specific stimulation of sensory and motor systems, produces a highly flexible and versatile network. The combinatorial power of this arrangement seems as though it could give nervous systems much of their range and responsiveness.

Interactions such as those just described are also likely to play a key role in how brains sort through the apparent chaos of their neural signaling (see Chapter 4). Because there is no one there to act as editor or to make summaries of the important findings, if the outcomes are to be adaptive for the individual, the system must have a means of biasing the waves of neuronal activity so that they are appropriate for the occasion.

Memories, Codes, and Patterns

A major role of value systems is to signal when something important is happening, such as alarming, harsh, or pleasant stimuli. Their role in learning is easy to see, such as when synapses are modified during training, but they may also be relevant to the experience of recognizing and remembering. There is an aspect to our own recognition and memory that involves the subjective feeling of having recaptured something. When memory is unreliable and we are certain of something, only to be proven completely wrong later, the experience surely involves the sense of having correctly remembered. Our brains are misleading us here, as if our value systems have been inappropriately activated by sensory events. Memories are, in some sense, the reliving of a previous experience. If the process of remembering involves the matching of patterns of neuronal activity, and if an inappropriate but similar enough pattern activates the value systems, a misperception will occur. In that case, memory will have failed.

No behavior is ever 100% consistent—not learning, memory recall, nor even fly courtship. Given what we have seen of the degeneracy of neuronal networks and mechanisms, the likelihood that recall will be performed by the brain in exactly the same way today as it was yesterday, let alone a week or a year ago, seems slim. When large sets of neurons are involved, which may be nearly all of the time, the occasional mismatch is inevitable. The resilience and flexibility of network degeneracy comes at the expense of accuracy. It seems to be our biological legacy that our brains trade off a narrow computer-like precision for a broader and more robust flexibility. Rigidity does not play very well out in nature.

When honeybees make the wrong choice on a memory test, perhaps it is because their brains are telling them that they are actually correct. (Or, alternatively, they are simply ignoring us and reacting indifferently.) If so, then we must wonder whether they are having the insect version of a subjective experience of feeling that they are right.

The issues of flexibility and occasional unreliability also raise the question of whether nervous systems use codes. Codes require a fair amount of precision, such as in the Morse code. The genetic code allows for a limited amount of degeneracy, but only to the extent that it stays within the strict limits of specifying the same amino acid. Accounting for how nervous systems make sense out of their chattering neurons may strain the idea of coding, especially given the large and variable number of cells and signals that accompany even the simplest responses, movements, and choices. The large number of cells active at a given time is well documented in *Aplysia* and the leech, and there is no reason to think that the same is not equally true for jellyfish and insects. A more flexible ability to recognize patterns, tempered by the activation of neuromodulators, may fit better with the capabilities and mechanisms of nervous systems. A flexible strategy also fits better with our evolutionary history, providing the elbow room to change and experiment without causing the whole system to crash.

Choice or Reflex?

In the early days of reductionist biology and psychology, all behavior was thought to be reflexive. The originator of this line of thinking, Jacques Loeb, studied the "tropisms" (orientation toward a stimulus) of various organisms, including caterpillars and cockroaches, and concluded that they were simply combinations of reflexes. Ivan Pavlov, the pioneer of associative conditioning (see Chapter 4), was heavily influenced by Loeb, and they agreed that essentially all behavior could be accounted for either by unconditioned (innate) or conditioned (learned) reflexes. This formulation left little room for choice or, at best, demoted choice to the level of alternative reflexes—for example, if presented with stimulus A, perform action P; if presented with stimulus B, perform action Q.

Many aspects of behavior certainly have the characteristics of reflex. *Paramecium*'s avoidance response, the barnacle's shadow reflex, *Aglantha*'s escape response, *Aplysia*'s gill withdrawal, and the fruit fly's haltere equilibrium compensation are several described in the foregoing chapters. Some behaviors, on the other hand, contain more complex elements than can easily be described as reflex. Search behavior is a good example, mate choice is another, and even the choice between swimming and crawling in the leech qualifies. The mix of factors that the animal's brain deals with, in such cases, includes not only immediate stimuli (of which there can be many), but also influences of the individual's past history on its nervous system. How the brain handles such a brew of activities to come out with an appropriate behavior is key to understanding the neural basis of decision-making.

Are Invertebrates Conscious?

The earliest studies of behavior in invertebrates viewed all of their actions as tropisms or reflexes. Such views followed in the footsteps of René Descartes, the 17th-century French philosopher and scientist who declared that only humans were conscious and all other creatures were automata. Ironically, it was the great *Paramecium* biologist H.S. Jennings in 1906 who first broached the idea of consciousness in organisms as simple as his little protozoans. Given what we know of the "neural" repertoire of a *Paramecium*, this now seems unlikely.

For animals such as the honeybee, however, with their demonstrated ability for learning abstract distinctions and maps, one may begin to wonder. Even the fruit fly may have something to contribute. Its tracking response includes the ability to anticipate and has many features that resemble attention, a trait generally associated with being conscious. The octopus is famous for its sophistication in solving problems. And jumping spiders have the capacity to plan a successful route through a maze simply by looking it over ahead of time.

If the capacity for consciousness depends on the complexity of circuitry in the brain rather than on one specific kind of anatomy, then the rudiments of consciousness (the neural strategies that are combined and built upon to make it possible) may be present in some invertebrates. Although these animals are certainly not conscious in the same way that humans are, they may offer accessible ways to study some components of consciousness. The spectrum of brains found in the invertebrate world certainly offers the richest set of examples for understanding how complex cognitive functions arise. If this seems far-fetched, it may be prudent to recall that the pioneers of molecular biology in the 1940s were laughed at for using bacterial viruses to study the nature of genes. They were told that such simple creatures could not possibly contribute anything useful to our understanding of "real" organisms, such as humans. The rest is history.

In the end, perhaps we will be able to resolve some of the most profound questions about our own brains by referring to universal principles that will be derived from some of the most distant organisms.

Further Reading

Bullock T.H. 2003. Have brain dynamics evolved? Should we look for unique dynamics in the sapient species? *Neural Comput.* **15:** 2013–2027.

Edelman G.M. 1998. Building a picture of the brain. *Daedalus* **127:** 37–69.

Greenspan R.J. and van Swinderen B. 2004. Cognitive consonance: Complex brain functions in the fruit fly and its relatives. *Trends Neurosci.* **27:** 707–711.

Koch C. 2004. *The quest for consciousness: A neurobiological approach*. Roberts & Company, Englewood, Colorado.

North G. and Greenspan R.J., eds. 2007. *Invertebrate neurobiology*. Cold Spring Harbor Laboratory Press, Cold Spring Harbor, New York. (In press).

Bibliography

Chapter 1

Baldauf S.L., Roger A.J., Wenk-Siefert I., and Doolittle W.F. 2002. A kingdom-level phylogeny of eukaryotes based on combined protein data. *Science* **290:** 972–977.

Blount P., Sukharev S.I., Moe P.C., Schroeder M.J., Guy H.R., and Kung C. 1996. Membrane topology and multimeric structure of a mechanosensitive channel protein of *Escherichia coli*. *EMBO J.* **15:** 4798–4805.

Chang S.-Y. and Kung C. 1973. Genetic analyses of heat-sensitive pawn mutants of *Paramecium aurelia*. *Genetics* **75:** 49–59.

Clark K.D., Hennessey T.M., Nelson D.F., and Preston R.R. 1997. Extracellular GTP causes membrane-potential oscillations through the parallel activation of Mg^{+2} and Na^+ currents in *Paramecium tetraurelia*. *J. Membr. Biol.* **157:** 159–167.

Doyle D.A., Morais-Cabral J., Pfuetzner R.A., Kuo A., Gulbis J.M., Cohen S.L., Chait B.T., and MacKinnon R. 1998. The structure of the potassium channel: Molecular basis of K^+ conduction and selectivity. *Science* **280:** 69–77.

Eckert R. 1972. Bioelectric control of ciliary activity. *Science* **176:** 473–481.

Eckert R., Naitoh Y., and Friedman K. 1972. Sensory mechanisms in *Paramecium*. I. Two components of the electric response to mechanical stimulation of the anterior surface. *J. Exp. Biol.* **56:** 683–694.

Edelman G.M. and Gally J.A. 2001. Degeneracy and complexity in biological systems. *Proc. Natl. Acad. Sci.* **98:** 13763–13768.

Haynes W.J., Ling K.Y., Saimi Y., and Kung C. 2003. PAK paradox: *Paramecium* appears to have more K^+-channel genes than humans. *Eukaryot. Cell* **2:** 737–745.

Kloda A. and Martinac B. 2002. Common evolutionary origins of mechanosensitive ion channels in Archaea, Bacteria and cell-walled Eukarya. *Archaea* **1:** 35–44.

Levina N., Totemeyer S., Stokes N.R., Louis P., Jones M.A., and Booth I.R. 1999. Protection of *Escherichia coli* cells against extreme turgor by activation of MscS and MscL mechanosensitive channels: Identification of genes required for MscS activity. *EMBO J.* **18:** 1730–1737.

Loukin S.H., Kuo M.M., Zhou X.L., Haynes W.J., Kung C., and Saimi Y. 2005. Microbial K^+ channels. *J. Gen. Physiol.* **125:** 521–527.

Martinac B. 2004. Mechanosensitive ion channels: Molecules of mechanotransduction. *J. Cell Sci.* **117:** 2449–2460.

Moe P.C., Blount P., and Kung C. 1998. Functional and structural conservation in the mechanosensitive channel MscL implicates elements crucial for mechanosensation. *Mol. Microbiol.* **28:** 583–592.

Naitoh Y. and Kaneko H. 1972. Reactivated triton-extracted models of *Paramecium*: Modification of ciliary movement by calcium ions. *Science* **176:** 523–524.

Ogura A. and Machemer H. 1980. Distribution of mechanoreceptor channels in the *Paramecium* surface membrane. *J. Comp. Physiol.* **135:** 233–242.

Schumacher M. and Adelman J.P. 2002. Ion channels: An open and shut case. *Nature* **417:** 501–502.

Zhou Y., Morais-Cabral J.H., Kaufman A., and MacKinnon R. 2001. Chemistry of ion coordination and hydration revealed by a K^+ channel-Fab complex at 2.0 Å resolution. *Nature* **414:** 43–48.

Chapter 2

Arendt D. and Wittbrodt J. 2001. Reconstructing the eyes of *Urbilateria*. *Philos. Trans. R. Soc. Lond. B* **356:** 1545–1563.

Bacigalupo J., Johnson E.C., Vergara C., and Lisman J.E. 1991. Light-dependent channels from excised patches of *Limulus* ventral photoreceptors are opened by cGMP. *Proc. Natl. Acad. Sci.* **88:** 7938–7942.

Callaway J.C. and Stuart A.E. 1989. Biochemical and physiological evidence that histamine is the transmitter of barnacle photoreceptors. *Vis. Neurosci.* **3:** 311–325.

Clark A.W., Millecchia R., and Mauro A. 1969. The ventral photoreceptor cells of *Limulus*. I. The microanatomy. *J Gen. Physiol.* **54:** 289–309.

Crisp D.J. and Southward A.J. 1961. Different types of cirral activity of barnacles. *Philos. Trans. R. Soc. Lond. B* **243:** 271–307.

Darwin C.R. 1851. *Living Cirripedia, Vol. 1: A monograph on the subclass Cirripedia, with figures of all the species. The Lepadidae; or, pedunculated cirripedes.* The Ray Society, London.

Dorlochter M. and Stieve H. 1997. The *Limulus* ventral photoreceptor: Light response and the role of calcium in a classic preparation. *Prog. Neurobiol.* **53:** 451–515.

Fein A. 2003. Inositol 1,4,5-trisphosphate-induced calcium release is necessary for generating the entire light response of *Limulus* ventral photoreceptors. *J. Gen. Physiol.* **121:** 441–449.

Garger A.V., Richard E.A., and Lisman J.E. 2004. The excitation cascade of *Limulus* ventral photoreceptors: Guanylate cyclase as the link between $InsP_3$-mediated Ca^{2+} release and the opening of cGMP-gated channels. *BMC Neurosci.* **5:** 7.

Garm A., Ekstrom P., Boudes M., and Nilsson D.E. 2006. Rhopalia are integrated parts of the central nervous system in box jellyfish. *Cell Tissue Res.* **325:** 333–343.

Hadrys T., DeSalle R., Sagasser S., Fischer N., and Schierwater B. 2005. The Trichoplax *PaxB* gene: A putative Proto-*PaxA/B/C* gene predating the origin of nerve and sensory cells. *Mol. Biol. Evol.* **22:** 1569–1578.

Hayashi J.H., Moore J.W., and Stuart A.E. 1985. Adaptation in the input–output relation of the synapse made by the barnacle's photoreceptor. *J. Physiol.* **368:** 179–195.

Hudspeth A.J. and Stuart A.E. 1977. Morphology and responses to light of the somata, axons, and terminal regions of individual photoreceptors of the giant barnacle. *J. Physiol.* **272:** 1–23.

Hudspeth A.J., Poo M.M., and Stuart A.E. 1977. Passive signal propagation and membrane properties in median photoreceptors of the giant barnacle. *J. Physiol.* **272:** 25–43.

Kozmik Z., Daube M., Frei E., Norman B., Kos L., Dishaw L.J., Noll M., and Piatigorsky J. 2003. Role of Pax genes in eye evolution: A cnidarian *PaxB* gene uniting Pax2 and Pax6 functions. *Dev. Cell* **5:** 773–785.

Leys S.P. and Degnan B.M. 2001. Cytological basis of photoresponsive behavior in a sponge larva. *Biol. Bull.* **201:** 323–338.

Li L. and Chin L.-S. 2003. The molecular machinery of synaptic vesicle exocytosis. *Cell Mol. Life Sci.* **60:** 942–960.

Millecchia R. and Gwilliam G.F. 1972. Photoreception in a barnacle: Electrophysiology of the shadow reflex pathway in *Balanus cariosus*. *Science* **177:** 438–440.

Nilsson D.E., Gislen L., Coates M.M., Skogh C., and Garm A. 2005. Advanced optics in a jellyfish eye. *Nature* **435:** 201–205.

Ozawa S., Hagiwara S., and Nicolaysen K. 1977. Neural organization of shadow reflex in a giant barnacle, *Balanus nubilus*. *J. Neurophysiol.* **40:** 982–995.

Piatigorsky J. and Kozmik Z. 2004. Cubozoan jellyfish: An Evo/Devo model for eyes and other sensory systems. *Int. J. Dev. Biol.* **48:** 719–729.

Chapter 3

Agassiz L. 1850. Contributions to the natural history of the acalephae of North America. Part I. On the naked-eyed medusae of the shores of Massachusetts, in their perfect state of development. *Mem. Amer. Acad. Arts Sci.,* new ser. **IV(part 2):** 221–316.

Anderson P.A. and Greenberg R.M. 2001. Phylogeny of ion channels: Clues to structure and function. *Comp. Biochem. Physiol. Part B Biochem. Mol. Biol.* **129:** 17–28.

Arkett S.A. and Mackie G.O. 1988. Hair cell mechanoreception in the jellyfish *Aglantha digitale*. *J. Exp. Biol.* **135:** 329–342.

Friesen W.O., Poon M., and Stent G.S. 1976. An oscillatory neuronal circuit generating a locomotory rhythm. *Proc. Natl. Acad. Sci.* **73:** 3734–3738.

Gorman A.L.F., Hermann A., and Thomas M.V. 1981. Intracellular calcium and the control of neuronal pacemaker activity. *Fed. Proc.* **40:** 2233–2239.

Hou X.-G., Aldridge R.J., Bengstrom J., Siveter D.J., and Feng X.-H. 2004. *The Cambrian fossils of Chengjang, China: The flowering of early animal life.* Blackwell Science Ltd., Oxford.

Kristan W., Calabrese R.L., and Friesen W.O. 2005. Neuronal control of leech behavior. *Prog. Neurobiol.* **76:** 279–327.

Kristan W.B., Stent G.S., and Ort C.A. 1974. Neuronal control of swimming in the medicinal leech. I. Dynamics of the swimming rhythm. *J. Comp. Physiol.* **94:** 97–119.

Mackie G.O. 1980. Slow swimming and cyclical "fishing" behavior in *Aglantha digitale* (Hydromedusae: Trachylina). *Can. J. Fish. Aquat. Sci.* **37:** 1550–1556.

——. 2004. Central neural circuitry in the jellyfish *Aglantha*: A model "simple nervous system." *Neurosignals* **13:** 5–19.

——. 2004. The first description of nerves in a cnidarian: Louis Agassiz's account of 1850. *Hydrobiologia* **530/531:** 27–32.

Mackie G.O. and Meech R.W. 2000. Central circuitry in the jellyfish *Aglantha digitale*. III. The rootlet and pacemaker systems. *J. Exp. Biol.* **203:** 1797–1807.

——. 1993. Ionic currents in giant motor axons of the jellyfish, *Aglantha digitale*. *J. Neurophys.* **69:** 884–893.

——. 1993. Potassium channel family in giant motor axons of *Aglantha digitale*. *J. Neurophys.* **69:** 894–901.

——. 1995. Synaptic potentials and threshold currents underlying spike production in motor giant axons of *Aglantha digitale*. *J. Neurophys.* **74:** 1662–1670.

Moroz L.L., Meech R.W., Sweedler J.V., and Mackie G.O. 2004. Nitric oxide regulates swimming in the jellyfish *Aglantha digitale*. *J. Comp. Neurol.* **471:** 26–36.

Peterson K.J. and Eernisse D.J. 2001. Animal phylogeny and the ancestry of bilaterians: Inferences from morphology and 18S rDNA gene sequences. *Evol. Dev.* **3:** 170–205.

Roberts A. and Mackie G.O. 1980. The giant axon escape system of a hydrozoan medusa, *Aglantha digitale*. *J. Exp. Biol.* **84:** 303–318.

Stent G.S., Kristan W.B., Friesen W.O., Ort C.A., Poon M., and Calabrese R.L. 1978. Neuronal generation of the leech swimming movement. *Science* **200:** 1348–1357.

Technau U., Rudd S., Maxwell P., Gordon P.M., Saina M., Grasso L.C., Hayward D.C., Sensen C.W., Saint R., Holstein T.W., Ball E.E., and Miller D.J. 2005. Maintenance of ancestral complexity and non-metazoan genes in two basal cnidarians. *Trends Genet.* **21:** 633–639.

Chapter 4

Briggman K.L., Abarbanel H.D.I., and Kristan W.B. 2005. Optical imaging of neuronal populations during decision-making. *Science* **307:** 896–901.

Byrne J.H. and Kandel E.R. 1996. Presynaptic facilitation revisited: State and time dependence. *J. Neurosci.* **16:** 425–435.

Carew T.J., Walters E.T., and Kandel E.R. 1981. Associative learning in *Aplysia*: Cellular correlates supporting a conditioned fear hypothesis. *Science* **211:** 501–504.

Erixon N.J., Demartini L.J., and Wright W.G. 1999. Dissociation between sensitization and learning-related neuromodulation in an aplysiid species. *J. Comp. Neurol.* **408:** 506–514.

Kicklighter C.E., Shabani S., Johnson P.M., and Derby C.D. 2005. Sea hares use novel antipredatory chemical defenses. *Curr. Biol.* **15:** 549–554.

Lewin M.R. and Walters E.T. 1999. Cyclic GMP pathway is critical for inducing long-term sensitization of nociceptive sensory neurons. *Nat. Neurosci.* **2:** 18–23.

Mackey S.L., Kandel E.R., and Hawkins R.D. 1989. Identified serotonergic neurons LCB1 and RCB1 in the cerebral ganglia of *Aplysia* produce presynaptic facilitation of siphon sensory neurons. *J. Neurosci.* **9:** 4227–4235.

Marinesco S., Duran K.L., and Wright W.G. 2003. Evolution of learning in three aplysiid species: Differences in heterosynaptic plasticity contrast with conservation in serotonergic pathways. *J. Physiol.* **550:** 241–253.

Marinesco S., Kolkman K.E., and Carew T.J. 2004. Serotonergic modulation in *Aplysia*. I. Distributed serotonergic network persistently activated by sensitizing stimuli. *J. Neurophysiol.* **92:** 2468–2486.

Shuster M.J., Camardo J.S., Siegelbaum S.A., and Kandel E.R. 1985. Cyclic AMP-dependent protein kinase closes the serotonin-sensitive K^+ channels of *Aplysia* sensory neurones in cell-free membrane patches. *Nature* **313:** 392–395.

Sutton M.A., Bagnall M.W., Sharma S.K., Shobe J., and Carew T.J. 2004. Intermediate-term memory for site-specific sensitization in *Aplysia* is maintained by persistent activation of protein kinase C. *J. Neurosci.* **24:** 3600–3609.

Trudeau L.E. and Castellucci V.F. 1993. Sensitization of the gill and siphon withdrawal reflex of *Aplysia*: Multiple sites of change in the neuronal network. *J. Neurophysiol.* **70:** 1210–1220.

Tsau Y., Wu J.-Y., Hopp H.-P., Cohen L.B., Schiminovich D., and Falk C.X. 1994. Distributed aspects of the response to siphon touch in *Aplysia*: Spread of stimulus information and cross-correlation analysis. *J. Neurosci.* **14:** 4167–4184.

Wright W.G. 1998. Evolution of nonassociative learning: Behavioral analysis of a phylogenetic lesion. *Neurobiol. Learn. Mem.* **69:** 326–337.

———. 2000. Neuronal and behavioral plasticity in evolution: Experiments in a model lineage. *Biosciences* **50:** 883–894.

Wu J.-Y., Tsau Y., Hopp H.-P., Cohen L.B., Tang A.C., and Falk C.X. 1994. Consistency in nervous systems: Trial-to-trial and animal-to-animal variations in the responses to repeated applications of a sensory stimulus in *Aplysia*. *J. Neurosci.* **14:** 1366–1384.

Zecevic D., Wu J.-Y., Cohen L.B., London J.A., Hopp H.P., and Falk C.X. 1989. Hundreds of neurons in the *Aplysia* abdominal ganglion are active during the gill-withdrawal reflex. *J. Neurosci.* **9:** 3681–3689.

Zhang H., Wainwright M., Byrne J.H., and Cleary L.J. 2003. Quantitation of contacts among sensory, motor, and serotonergic neurons in the pedal ganglion of *Aplysia*. *Learn. Mem.* **10:** 387–393.

Chapter 5

Allada R., White N.E., So W.V., Hall J.C., and Rosbash M. 1998. A mutant *Drosophila* homolog of mammalian *Clock* disrupts circadian rhythms and transcription of *period* and *timeless*. *Cell* **93:** 791–804.

Costa R., Peixoto A.A., Barbujani G., and Kyriacou C.P. 1992. A latitudinal cline in a *Drosophila* clock gene. *Proc. Biol. Sci.* **250:** 43–49.

Grima B., Chelot E., Xia R., and Rouyer F. 2004. Morning and evening peaks of activity rely on different clock neurons of the *Drosophila* brain. *Nature* **431:** 869–873.

Hardin P.E., Hall J.C., and Rosbash M. 1992. Circadian oscillations in *period* gene mRNA levels are transcriptionally regulated. *Proc. Natl. Acad. Sci.* **89:** 11711–11715.

Hyun S., Lee Y., Hong S.T., Bang S., Paik D., Kang J., Shin J., Lee J., Jeon K., Hwang S., Bae E., and Kim J. 2005. *Drosophila* GPCR Han is a receptor for the circadian clock neuropeptide PDF. *Neuron* **48:** 267–278.

Konopka R.J. and Benzer S. 1971. Clock mutants of *Drosophila melanogaster*. *Proc. Nat. Acad. Sci.* **68:** 2112–2116.

Kyriacou C.P. and Hall J.C. 1980. Circadian rhythm mutations in *Drosophila* affect short-term fluctuations in the male's courtship song. *Proc. Natl. Acad. Sci.* **77:** 6729–6733.

———. 1982. The function of courtship song rhythms in *Drosophila*. *Anim. Behav.* **30:** 794–801.

Lear B.C., Merrill C.E., Lin J.M., Schroeder A., Zhang L., and Allada R. 2005. A G protein-coupled receptor, groom-of-PDF, is required for PDF neuron action in circadian behavior. *Neuron* **48:** 221–227.

Lin Y., Stormo G.D., and Taghert P.H. 2004. The neuropeptide Pigment-Dispering Factor coordinates pacemaker interaction in the *Drosophila* nervous system. *J. Neurosci.* **24:** 7951–7957.

Nitabach M.N., Blau J., and Holmes T.C. 2002. Electrical silencing of *Drosophila* pacemaker neurons stops the free-running circadian clock. *Cell* **109:** 485–495.

Nitz D.A., van Swinderen B., Tononi G., and Greenspan R.J. 2002. Electrophysiological correlates of rest and activity in *Drosophila melanogaster*. *Curr. Biol.* **12:** 1934–1940.

Price J.L., Blau J., Rothenfluh A., Abodeely M., Kloss B., and Young M.W. 1998. *double-time* is a novel *Drosophila* clock gene that regulates PERIOD protein accumulation. *Cell* **94:** 83–95.

Renn S.C., Park J.H., Rosbash M., Hall J.C., and Taghert P.H. 1999. A *pdf* neuropeptide gene mutation and ablation of PDF neurons each cause severe abnormalities of behavioral circadian rhythms in *Drosophila*. *Cell* **99:** 791–802.

Rutila J.E., Suri V., Le M., So W.V., Rosbash M., and Hall J.C. 1998. CYCLE is a second bHLH-PAS clock protein essential for circadian rhythmicity and transcription of *Drosophila period* and *timeless. Cell* **93:** 805–814.

Sawyer L.A., Hennessy J.M., Peixoto A.A., Rosato E., Parkinson H., Costa R., and Kyriacou C.P. 1997. Natural variation in a *Drosophila* clock gene and temperature compensation. *Science* **278:** 2117–2120.

Schopf J.W. 1994. Disparate rates, differing fates: Tempo and mode of evolution changed from the Precambrian to the Phanerozoic. *Proc. Natl. Acad. Sci.* **91:** 6735–6742.

Sehgal A., Price J.L., Man B., and Young M.W. 1994. Loss of circadian behavioral rhythms and per RNA oscillations in the *Drosophila* mutant *timeless. Science* **263:** 1603–1606.

Shaw P.J., Cirelli C., Greenspan R.J., and Tononi G. 2000. Correlates of sleep and waking in *Drosophila melanogaster. Science* **287:** 1834–1837.

Stanewsky R., Kaneko M., Emery P., Beretta B., Wager-Smith K., Kay S.A., Rosbash M., and Hall J.C. 1998. The cry^b mutation identifies cryptochrome as a circadian photoreceptor in *Drosophila. Cell* **95:** 681–692.

Stoleru D., Peng Y., Agosto J., and Rosbash M. 2004. Coupled oscillators control morning and evening locomotor behaviour of *Drosophila. Nature* **431:** 862–868.

Stoleru D., Peng Y., Nawathean P., and Rosbash M. 2005. A resetting signal between *Drosophila* pacemakers synchronizes morning and evening activity. *Nature* **438:** 238–242.

Zerr D.M., Hall J.C., Rosbash M., and Siwicki K.K. 1990. Circadian fluctuations of period protein immunoreactivity in the CNS and the visual system of *Drosophila. J. Neurosci.* **10:** 2749–2762.

Chapter 6

Andretic R., van Swinderen B., and Greenspan R.J. 2005. Dopaminergic modulation of arousal in *Drosophila. Curr. Biol.* **15:** 1165–1175.

Borst A. and Haag J. 2002. Neural networks in the cockpit of the fly. *J. Comp. Physiol. A* **188:** 419–437.

Broughton S.J., Kitamoto T., and Greenspan R.J. 2004. Excitatory and inhibitory switches for courtship in the brain of *Drosophila melanogaster. Curr. Biol.* **14:** 538–547.

Chan W.P., Prete F., and Dickinson M.H. 1998. Visual input to the efferent control system of a fly's "gyroscope." *Science* **280:** 289–292.

Coyne J.A., Boussy I.A., Prout T., Bryant S.H., and Jones J.S. 1982. Long-distance migration of *Drosophila. Am. Nat.* **119:** 589–595.

Dickinson M.H. 1999. Haltere-mediated equilibrium reflexes of the fruit fly, *Drosophila melanogaster. Philos. Trans. R. Soc. Lond. B* **354:** 903–916.

Egelhaaf M. 1985. On the neuronal basis of figure-ground discrimination by relative motion in the visual system of the fly. *Biol. Cybernet.* **52:** 195–209.

Engel J.E. and Wu C.-F. 1992. Interactions of membrane excitability mutations affecting potassium and sodium currents in the flight and giant fiber escape systems of *Drosophila. J. Comp. Physiol. A* **171:** 93–104.

Fayyazuddin A. and Dickinson M.H. 1999. Convergent mechanosensory input structures the firing phase of a steering motor neuron in the blowfly, *Calliphora. J. Neurophysiol.* **82:** 1916–1926.

Frye M.A., Tarsitano M., and Dickinson M.H. 2003. Odor localization requires visual feedback during free flight in *Drosophila melanogaster. J. Exp. Biol.* **206:** 843–855.

Götz K. 1975. The optomotor equilibrium of the *Drosophila* navigation system. *J. Comp. Physiol. A* **99:** 187–210.

Guo A. and Götz K.G. 1997. Association of visual objects and olfactory cues in *Drosophila*. *Learn. Mem.* **4:** 192–204.

Harcombe E.S. and Wyman R.J. 1978. The cyclically repetitive firing sequences of identified *Drosophila* flight motoneuron. *J. Comp. Physiol. A* **123:** 271–279.

King D.G. and Valentino K.L. 1983. On neuronal homology: A comparison of similar axons in *Musca, Sarcophaga,* and *Drosophila* (Diptera, Schizophora). *J. Comp. Neurol.* **219:** 1–9.

Krapp H.G. and Hengstenberg R. 1996. Estimation of self-motion by optic flow processing in single visual interneurons. *Nature* **384:** 463–466.

Sherman A. and Dickinson M.H. 2003. A comparison of visual and haltere-mediated equilibrium reflexes in the fruit fly *Drosophila melanogaster*. *J. Exp. Biol.* **206:** 295–302.

Trimarchi J.R. and Schneiderman A.M. 1995. Different neural pathways coordinate *Drosophila* flight initiations evoked by visual and olfactory stimuli. *J. Exp. Biol.* **198:** 1099–1104.

———. 1995. Flight initiations in *Drosophila melanogaster* are mediated by several distinct motor patterns. *J. Comp. Physiol. A* **176:** 355–364.

———. 1995. Initiation of flight in the unrestrained fly, *Drosophila melanogaster*. *J. Zool.* **235:** 211–222.

van Swinderen B. and Greenspan R.J. 2003. Salience modulates 20–30 Hz brain activity in *Drosophila*. *Nat. Neurosci.* **6:** 579–586.

Chapter 7

Andretic R., van Swinderen B., and Greenspan R.J. 2005. Dopaminergic modulation of arousal in *Drosophila*. *Curr. Biol.* **15:** 1165–1175.

Billeter J.C. and Goodwin S.F. 2004. Characterization of *Drosophila fruitless-gal4* transgenes reveals expression in male-specific *fruitless* neurons and innervation of male reproductive structures. *J. Comp. Neurol.* **475:** 270–287.

Bray S. and Amrein H. 2003. A putative *Drosophila* pheromone receptor expressed in male-specific taste neurons is required for efficient courtship. *Neuron* **39:** 1019–1029.

Campesan S., Dubrova Y., Hall J.C., and Kyriacou C.P. 2001. The *nonA* gene in *Drosophila* conveys species-specific behavioral characteristics. *Genetics* **158:** 1535–1543.

Chen P.S., Stumm-Zollinger E., Aigaki T., Balmer J., Bienz M., and Böhlen P. 1988. A male acessory gland peptid that regulates reproductive behavior of female *Drosophila melanogaster*. *Cell* **54:** 291–298.

Coyne J.A. and Charlesworth B. 1997. Genetics of a pheromonal difference affecting sexual isolation between *Drosophila mauritiana* and *D. sechellia*. *Genetics* **145:** 1015–1030.

Demir E. and Dickson B.J. 2005. *fruitless* splicing specifies male courtship behavior in *Drosophila*. *Cell* **121:** 785–794.

Ewing A.W. 1979. The neuromuscular basis of courtship song in *Drosophila*: The role of the direct and axillary wing muscles. *J. Comp. Physiol.* **130:** 87–93.

Ferveur J.-F., Cobb M., Boukella H., and Jallon J.M. 1996. World-wide variation in *Drosophila melanogaster* sex pheromone: Behavioural effects, genetic bases and potential evolutionary consequences. *Genetica* **97:** 73–80.

Ferveur J.-F., Störtkuhl K., Stocker R.F., and Greenspan R.J. 1995. Genetic feminization of brain structures and changed sexual orientation in male *Drosophila melanogaster*. *Science* **267:** 902–905.

Hall J.C. 1979. Control of male reproductive behavior by the central nervous system of *Drosophila*: Dissection of a courtship pathway by genetic mosaics. *Genetics* **92:** 437–457.

Kulkarni S.J., Steinlauf A.F., and Hall J.C. 1988. The *dissonance* mutant of courtship song in *Drosophila melanogaster*: Isolation, behavior and cytogenetics. *Genetics* **118:** 267–285.

Kyriacou C.P. and Hall J.C. 1980. Circadian rhythm mutations in *Drosophila melanogaster* affect short-term fluctuations in the male's courtship song. *Proc. Natl. Acad. Sci.* **77:** 6729–6733.

———. 1985. Action potential mutations stop a biological clock in *Drosophila*. *Nature* **314:** 171–173.

Manoli D.S. and Baker B.S. 2004. Median bundle neurons coordinate behaviours during *Drosophila* male courtship. *Nature* **430:** 564–569.

Manoli D.S., Foss M., Villella A., Taylor B.J., Hall J.C., and Baker B.S. 2005. Male-specific *fruitless* specifies the neural substrates of *Drosophila* courtship behaviour. *Nature* **436:** 395–400.

Tauber E., Roe H., Costa R., Hennessy J.M., and Kyriacou C.P. 2003. Temporal mating isolation driven by a behavioral gene in *Drosophila*. *Curr. Biol.* **13:** 140–145.

von Schilcher F. 1976. The role of auditory stimuli in the courtship of *Drosophila melanogaster*. *Anim. Behav.* **24:** 18–26.

von Schilcher F. and Hall J.C. 1979. Neural topography of courtship song in sex mosaics of *Drosophila melanogaster*. *J. Comp. Physiol. A* **129:** 85–95.

Wheeler D.A., Kyriacou C.P., Greenacre M.L., Yu Q., Rutila J.E., Rosbash M., and Hall J.C. 1991. Molecular transfer of a species-specific behavior from *Drosophila simulans* to *Drosophila melanogaster*. *Science* **251:** 1082–1085.

Chapter 8

Dubnau J., Chiang A.S., Grady L., Barditch J., Gossweiler S., McNeil J., Smith P., Buldoc F., Scott R., Certa U., Broger C., and Tully T. 2003. The *staufen/pumilio* pathway is involved in *Drosophila* long-term memory. *Curr. Biol.* **13:** 286–296.

Faber T., Joerges J., and Menzel R. 1999. Associative learning modifies neural representations of odors in the insect brain. *Nat. Neurosci.* **2:** 74–78.

Giurfa M., Eichmann B., and Menzel R. 1996. Symmetry perception in an insect. *Nature* **382:** 458–461.

Giurfa M., Zhang S., Jenett A., Menzel R., and Srinivasan M.V. 2001. The concepts of "sameness" and "difference" in an insect. *Nature* **410:** 930–933.

Griffith L.C., Verselis L.M., Aitken K.M., Kyriacou C., Danho W., and Greenspan R.J. 1993. Inhibition of calcium/calmodulin-dependent protein kinase in *Drosophila* disrupts behavioral plasticity. *Neuron* **10:** 501–509.

Hammer M. 1993. An identified neuron mediates the unconditioned stimulus in associative olfactory learning in honeybees. *Nature* **366:** 59–63.

Hammer M. and Menzel R. 1998. Multiple sites of associative odor learning as revealed by local brain microinjections of octopamine in honeybees. *Learn. Mem.* **5:** 146–156.

Joiner M.L. and Griffith L.C. 1999. Mapping of the anatomical circuit of CaM kinase-dependent courtship conditioning in *Drosophila*. *Learn. Mem.* **6:** 177–192.

Kim Y.K., Koepfer H.R., and Ehrman L. 1996. Developmental isolation and subsequent adult behavior of *Drosophila paulistorum*. *Behav. Genet.* **26:** 27–37.

Kim Y.K., Phillips D.R., Chao T., and Ehrman L. 2004. Developmental isolation and subsequent adult behavior of *Drosophila paulistorum*. VI. Quantitative variation in cuticular hydrocarbons. *Behav. Genet.* **34:** 385–394.

Menzel R., Manz G., Menzel R., and Greggers U. 2001. Massed and spaced learning in honeybees: The role of CS, US, the intertrial interval, and the test interval. *Learn. Mem.* **8:** 198–208.

Menzel R., Geiger K., Müller U., Joerges J., and Chittka L. 1998. Bees travel novel homeward routes by integrating separately acquired vector memories. *Anim. Behav.* **55:** 139–152.

Menzel R., Greggers U., Smith A., Berger S., Brandt R., Brunke S., Bundrock G., Hulse S., Plumpe T., Schaupp F., Schuttler E., Stach S., Stindt J., Stollhoff N., and Watzl S. 2005. Honey bees navigate according to a map-like spatial memory. *Proc. Natl. Acad. Sci.* **102:** 3040–3045.

Mizunami M., Weibrecht J.M., and Strausfeld N.J. 1998. Mushroom bodies of the cockroach: Their participation in place memory. *J. Comp. Neurol.* **402:** 520–537.

Mizunami M., Okada R., Li Y., and Strausfeld N.J. 1998. Mushroom bodies of the cockroach: Activity and identities of neurons recorded in freely moving animals. *J. Comp. Neurol.* **402:** 501–519.

Sandoz J.C., Galizia C.G., and Menzel R. 2003. Side-specific olfactory conditioning leads to more specific odor representation between sides but not within sides in the honeybee antennal lobes. *Neuroscience* **120:** 1137–1148.

Savarit F., Sureau G., Cobb M., and Ferveur J.F. 1999. Genetic elimination of known pheromones reveals the fundamental chemical bases of mating and isolation in *Drosophila*. *Proc. Natl. Acad. Sci.* **96:** 9015–9020.

Siegel R.W. and Hall J.C. 1979. Conditioned responses in courtship behavior of normal and mutant *Drosophila*. *Proc. Natl. Acad. Sci.* **76:** 3430–3434.

Stocker R.F. and Gendre N. 1989. Courtship behavior of *Drosophila*, genetically and surgically deprived of basiconic sensilla. *Behav. Genet.* **19:** 371–385.

Tully T. and Quinn W.G. 1985. Classical conditioning and retention in normal and mutant *Drosophila melanogaster*. *J. Comp. Physiol. A* **157:** 263–277.

Tully T., Preat T., Boynton S.C., and Del Vecchio M. 1994. Genetic dissection of consolidated memory in *Drosophila*. *Cell* **79:** 35–47.

Yin J.C., Wallach J.S., Del Vecchio M., Wilder E.L., Zhou H., Quinn W.G., and Tully T. 1994. Induction of a dominant negative CREB transgene specifically blocks long-term memory in *Drosophila*. *Cell* **79:** 49–58.

Zawistowski S. and Richmond R.C. 1985. Experience-mediated courtship reduction and competition for mates by male *Drosophila melanogaster*. *Behav. Genet.* **15:** 561–569.

Chapter 9

Descartes R. 1985–1991. *The philosophical writings of Descartes*, 3 vols., translated by Cottingham J., Stoothoff R., and Murdoch D., volume 3 including A. Kenny. Cambridge University Press, Cambridge.

Edelman G.M. and Tononi G. 2000. *A universe of consciousness*. Basic Books, New York.

Greenspan R.J. and Baars B.J. 2004. Consciousness eclipsed: Jacques Loeb, Ivan P. Pavlov, and the triumph of reductionistic biology after 1900. *Consciousness Cognition* **14:** 220–231.

Jennings H.S. 1906. *Behavior of the lower organisms*. Columbia University Press, New York.

Loeb J. 1900. *Comparative physiology of the brain and comparative psychology*. G.P. Putnam's Sons, New York.

Pavlov I.P. 1927. *Conditioned reflexes and psychiatry. Lectures on Conditioned Reflexes*. Vols. 1 and 2, trans. and ed. W.H. Gantt, New York.

Glossary

abdominal ganglion posterior-most part of the fruit fly's nervous system

action potential the electrical signal that rapidly propagates along the axon of nerve cells as well as over the surface of muscle cells, because of a change in membrane potential caused by a flow of ions across the membrane through voltage-activated ion channels

affinity propensity for binding one molecule (protein, ligand, etc.) to another

allele a variant form of a gene; includes naturally occurring sequence variants, induced mutations, or any other alteration in a gene's sequence

allopatric occurring in separate, nonoverlapping geographic areas

archaea one of the three domains of the living world (bacteria, archaea, and eukarya), consisting of prokaryotic unicells with gene sequences that resemble those of eukaryotes more than those of bacteria, many of which live in extreme environments

arista the most distal part of the fruit fly antenna; a feather-like structure that moves in response to air currents or sound waves

associative conditioning type of learning that occurs when two stimuli are presented together (or nearly so) in time, one of which is neutral (the conditioned stimulus, CS), and the other of which would elicit a response on its own (the unconditioned stimulus, US). Once the training is complete, the animal gives the same response to the CS as it did originally to the US.

axon elongated extension of a neuron's membrane, terminating in a synapse. Action potentials are conducted along axon membranes.

Cambrian period in Earth's early history (543–490 million years ago) when animal life flowered and the basic phyla and body plans of all subsequent animal types first appeared

campaniform sensilla sensory structure on the fruit fly's exoskeleton that detects deformations (bending) of the surface; contains a sensory neuron with mechanoreceptors

central complex region of the insect brain found in the middle of the main lobe, the protocerebrum, implicated in control of locomotor behavior

cilia microscopic, hair-shaped extensions of the membrane on the surface of cells. On protozoans, they are capable of movement and mediate locomotor behavior.

Ciliate class of protozoans in the phylum Ciliophora, characterized by the presence of cilia on their surface

circadian daily or lasting for approximately 24 hours

cline a series of neighboring populations that exhibit gradual and continuous change of some character in response to an environmental gradient

compound eye a type of eye found in Arthropods that is composed of many light-sensitive units (ommatidia), each having its own lens

compound flower a flower head composed of multiple florets enclosed in a common leaf-like sheath (bract), such as a sunflower or dandelion

conformation the shape or arrangement of a molecule's structure or components

conservation evolutionary preservation of some trait, such as function, form, biochemistry, or gene sequence

critical period time period in the life cycle of an animal in which a specific kind of experience can modify the nervous system in a long-lasting or permanent manner

cryptochrome protein involved in circadian rhythms containing a flavin molecule as its light-absorbing structure

cyanobacteria prokaryote capable of photosynthesis (previously known as blue-green algae)

degeneracy system property in which different structures or arrangements produce the same output (in contrast to redundancy, where system structures are identical)

dendrite the specialized branches that extend from a neuron's cell body and function to receive messages from other neurons

depolarize to make the membrane potential less negative

domains the three major divisions of living organisms, bacteria, archaea, and eukarya, based on DNA sequence similarities

dopaminergic neurons that use dopamine as their neurotransmitter

dorsal anatomical designation for structures that are at or near the top of an organism's anatomy

endosymbiont an organism that lives inside another organism in a mutually beneficial relationship

epistemology the philosophical discipline that examines the nature of human knowledge and cognition

epithelial cells that line the internal and external surfaces of an organism's body

eukarya one of the three domains of living organisms, containing all organisms with cells that have a nucleus and organelles (i.e., all animals, plants, fungi, and protists)

flavin a class of coenzymes that occurs universally in living organisms and plays important roles in biochemical oxidations and reductions. It also occurs as the photopigment in the light-absorbing Cryptochrome protein involved in circadian rhythms.

floret small flowers that occur in groups on the same stem, making a compound flower

ganglion cluster of neurons that form an anatomically recognizable structure

glomeruli cluster of neurons characteristic of olfactory systems that receive synapses from sensory cells and send axons into the brain

gustatory pertaining to the sensory system specialized for taste

haltere a small knob-like balancing organ located below the wing in two-winged flies

homologous similar in appearance or structure and thus presumed to be derived from a common ancestor

hyperpolarize to make the membrane potential more negative

interneuron a neuron that communicates only to other neurons

ion channel protein in a cell's surface membrane that controls the flow of ions into the cell

kinase an enzyme that transfers phosphate groups

lamina outermost optic lobe of the insect nervous system

ligand a small molecule that binds specifically to a protein

lobula one of a pair of innermost optic lobes in the insect nervous system; the other is lobula plate

lobula plate one of a pair of innermost optic lobes in the insect nervous system; *see* lobula

medulla middle optic lobe, between lamina and lobula/lobula plate, in the insect nervous system

mesozoan the simplest animals, lacking a full metazoan organization

metazoan all animals having a body composed of cells differentiated into tissues and organs

microelectrode a very fine wire or hollow glass needle used to record electrical signals from nerve cells

microvilli minute projections of cell membranes that greatly increase the membranes' surface area

motor neuron neuron that synapses onto a muscle

mushroom bodies structures in the dorsal insect brain that are somewhat mushroom shaped and participate in olfaction, learning, and higher-order cognitive functions

myoepithelium sheet-like cells in jellyfish that have muscle contractile proteins but that are part of an epithelial sheet

neuromodulator class of neurotransmitters that modifies the activity of other neurons

neuropeptide a neurotransmitter that is a small sequence of amino acids

ommatidia hexagonal subunits of the compound eye found in Arthropods containing photoreceptor, pigment, and lens cells

opsin protein in photoreceptor cells that absorbs light using the small molecule retinal derived from vitamin A

osmolarity the total molar concentration of solutes in a liquid

osmotic the pressure differential that exists between two solutions with different concentrations of solutes when placed on opposite sides of a semipermeable membrane

passive spread movement of electrical current along the membrane of a neuron as if it were a cable, with decrement of the signal as it spreads

pheromone a chemical produced and secreted to signal from one individual to another of the same species

phosphorylate transfer of a phosphate group from one molecule to another, performed by a kinase

photopigment the light-absorbing small molecule in photosensitive proteins—retinal in rhodopsin and flavin in Cryptochrome

photoreceptor sensory cell specialized for responding to light

phylogenetic tree a diagram showing the evolutionary distance between organisms inferred from one or more characteristics (e.g., morphology, DNA sequence)

pitch the tilt up or down of the nose of a flying body

Placozoan the only phylum of mesozoans, the simplest of all animals lacking distinct organs and having few differentiated cell types

plasticity modifiability of physiological or anatomical properties of neurons as the result of prior experience

postsynaptic the cell or portion of membrane that is contacted by a synapse from which it receives neurotransmission

presynaptic the cell or portion of membrane that contacts another cell at a synapse and that releases neurotransmitter

proboscis an insect's mouth parts that can be extended and that have taste receptors on the end

Proterozoic the period of geologic time from 2500 to 543 million years ago, immediately preceding the Cambrian period

protists unicellular eukaryotes (*see* protozoan)

protocerebrum dorsal portion of the insect brain

protozoan unicellular eukaryotes (*see* protists)

quantum smallest unit, referring to light, to neurotransmission (corresponding to the release of a single vesicle), or to photoresponse (corresponding to the electrical response to light absorption by a single molecule of rhodopsin)

receptor protein that binds a ligand and subsequently initiates an intracellular signal

redundancy system property that allows it to continue functioning normally in the event of a component failure, by having backup components that perform duplicate functions (contrasts with degeneracy)

retinal small molecule in rhodopsin that absorbs light

rhodopsin combination of retinal and opsin

rhopalium sensory structure found in some jellyfish that contains photoreceptors, in simple and complex organization, as well as mechanoreceptors

salient having the quality of standing out and commanding attention

semispecies populations of a species that are completely isolated from one another but have not yet become truly different species

sensitization behavioral modification in which the response to one stimulus is heightened by previous exposure to another salient stimulus (often aversive)

serotonergic neurons using serotonin (5HT) as a neurotransmitter

spike *see* action potential

statocyst touch-sensitive organ on jellyfish that contains mechanoreceptors

sympatric occupying the same or overlapping geographic areas without interbreeding

synapse structure at the end of a neuron that contacts another cell and releases neurotransmitter

synaptic transmission the process of signaling between cells at a synapse whereby the presynaptic cell releases neurotransmitter that then binds to receptors on the postsynaptic cell and there initiates a signal

thoracic ganglion the portion of an insect's central nervous system in its thorax

threshold the level of membrane depolarization that must be exceeded in order to initiate an action potential

trace record of the change in membrane potential in a neuron

translational movement motion of visual stimuli that move past eyes on both sides in a front-to-back direction

transmembrane spanning the lipid bilayer, refers to proteins whose amino acid sequence is partially embedded in the membrane and protruding on both sides

transporter membrane protein that pumps ions or other small molecules into or out of the cell

ultradian rhythms that occur more than once per day

ventral anatomical designation for structures that are at or near the bottom of a structure

Credits

Abbreviations: AAAS, American Association for the Advancement of Science; NAS, USA, National Academy of Sciences, U.S.A.; NOAA, National Oceanic & Atmospheric Administration; CNRC, National Research Council Canada; FASEB, Federation of American Societies for Biology; CSHLP, Cold Spring Harbor Laboratory Press.

Chapter 1

Quote, reprinted from Jennings H.S., *Behavior of the lower organisms,* © 1915 Columbia University Press; **1.1,** with permission from Rijksmuseum, Amsterdam; **1.2,** redrawn from Jennings H.S., *Behavior of the lower organisms,* © 1915 Columbia University Press; **1.3,** modified from Allen R.D., *J. Cell Biol. 49:* 1–20, © 1971 Rockefeller University Press, and Jennings H.S., *Behavior of the lower organisms,* © 1915 Columbia University Press; **1.4 and 1.7,** redrawn from Eckert R., *Science 176:* 473–481, © 1972 AAAS; **1.5,** modified from Eckert R. et al., *J. Exp. Biol. 56:* 683–694, © 1972 Company of Biologists; **1.6,** modified from Naitoh Y et al., *Science 176:* 523–524, © 1972 AAAS; **1.8,** reprinted from Chang S.-Y. et al., *Genetics 75:* 49–59, © 1973 Genetics Society of America; **1.9,** modified from Ogura A. et al., *J. Comp. Physiol. 135:* 233–242, © 1980 Springer Science and Business Media; **1.10,** redrawn from Clark K.D. et al., *J. Membrane Biol. 157:* 159–167, © 1997 Springer Science and Business Media; **1.11,** modified from Pace N., *Science 276:* 734–740, © 2005 AAAS; **1.12,** modified from Doyle D.A. et al., *Science 280:* 69–77, © 1998 AAAS, and Zhou Y. et al., *Nature 414:* 43–48, © 2001 Macmillan; **1.13, top,** modified from Ahern C.A. et al., *J. Gen. Physiol. 123:* 205–216, © 2004 Rockefeller University Press; **1.13, bottom,** modified from Schumacher M. et al., *Nature 417:* 501–502, © 2002 Macmillan; **1.14,** modified from Kloda A. et al., *Archaea 1:* 35–44, © 2002 Heron Publishing, Victoria, British Columbia; **1.15,** modified from Perozo E. et al., *Nat. Struct. Biol. 9:* 696–703, © 2002 Macmillan; **1.16,** © 1995, Mitchell L. Sogin and David J. Patterson, Marine Biological Laboratory, Woods Hole, Massachusetts, with modifications.

Chapter 2

Quote and 2.1, reprinted from Darwin-online.org.uk, The Complete Work of Charles Darwin Online, University of Cambridge; **2.2,** modified from Crisp D.J. et al., *Philos. Trans. R. Soc. London B 243:* 271–307, © 1961 The Royal Society; **2.3,** modified from Hudspeth A.J. et al., *J. Physiol. 272:* 1–23, © 1977 Blackwell Publishing; **2.4,** modified from Hudspeth A.J. et al., *J. Physiol. 272:* 25–43, © 1977 Blackwell Publishing; **2.6,** modified from Ozawa S. et al., *J. Neurophysiol. 40:* 982–995, © 1977 with permission from the American Physiological Society; **2.7,** image provided by Genny Anderson, Santa Barbara City College, used with permission; **2.8,** modified from Millecchia R. et al.,

Science 177: 438–440, © 1972 AAAS; **2.9,** reprinted from Clark A.W., *J. Gen. Physiol. 54*: 289–309, © 1969 Rockefeller University Press; **2.13,** modified from Bacigalupo J. et al., *Proc. Natl. Acad. Sci. 88*: 7938–7942, © 1991 NAS, USA; **2.14,** modified from Li L. et al., *Cell Mol. Life Sci. 60*: 942–960, © 2003 Birkhauser Verlag, Switzerland; **2.16,** redrawn from Callaway J.C. et al., *Vis. Neurosci. 3*: 311–325, © 1989 Cambridge University Press; **2.17,** redrawn from Hayashi J.H. et al., *J. Physiol. 368*: 179–195, © 1985 Blackwell Publishing; **2.18,** modified from Arendt D. et al., *Philos. Trans. R. Soc. London B 356*: 1545–1563, © 2001 The Royal Society, Jensen B., *Iridology, Vol. 2*, p. 135, © 1982 Bernard Jensen Enterprises, Escondido, California, Land M.F. et al., *Animal eyes*, p. 128, © 2002 Oxford University Press, and Nilsson D.-E. *Facets of vision*, pp. 30–73, © 1989 Springer Science and Business Media; **2.19,** modified from Piatigorsky J. et al., *Int. J. Dev. Biol. 48*: 719–729, © 2004 UBC Press, Bilbao, Spain; **2.20,** courtesy of Dan-E. Nilsson, used with permission.

Chapter 3

Quote, reprinted from Eliot T.S., *Knowledge and experience in the philosophy of F.H. Bradley*, © 1964 by T.S. Eliot, Faber and Faber, London; **3.1, left,** image provided by Pamela J.W. Gore, Georgia Perimeter College, used with permission; **3.1, right,** reproduced from Hou X.-G. et al., *The Cambrian fossils of Chengjiang, China: The flowering of early animal life*, p. 57, © 2004 Blackwell Science Ltd., Malden, Massachusetts, image courtesy of Xianguang Hou; **3.2,** modified from Peterson K.J. et al., *Evol. Dev. 3*: 170–205, © 2001 Blackwell Publishing Ltd.; **3.3,** modified from Technau U. et al., *Trends Genet. 21*: 633–639, © 2005 Elsevier; **3.4, left,** NOAA; **3.4, right,** reproduced from Agassiz L., 1850, *Mem. Am. Acad. Arts Sci. 4*: 221–316, image courtesy of Prof. George Mackie, University of Victoria, British Columbia; **3.5,** modified from Moroz L.L. et al., *J. Comp. Neurol. 471*: 26–36, © 2004 John Wiley & Sons, Inc., Mackie G. et al., *J. Exp. Biol. 198 (Pt II)*: 2261–2270, © 1995 Company of Biologists, and Roberts A. et al., *J. Exp. Biol. 84*: 303–318, © 1980 Company of Biologists; **3.6,** redrawn from Mackie G.O., *Can. J. Fish Aquat. Sci. 37*: 1550–1556, © 1980 CNRC; **3.7 and 3.11,** modified with permission from Meech R.W. et al., *J. Neurophys. 74*: 1662–1670, © 1995 The American Physiological Society; **3.8,** modified from Mackie G.O. et al., *J. Exp. Biol. 203*: 1797–1807, © 2000 Company of Biologists; **3.9,** modified from Arkett S.A., *J. Exp. Biol. 135*: 329–342, © 1988 Company of Biologists; **3.10,** modified from Roberts A. et al., *J. Exp. Biol. 84*: 303–318, © 1980 Company of Biologists; **3.12,** redrawn with permission from Meech R.W. et al., *J. Neurophys. 69*: 884–893, © 1993 The American Physiological Society; **3.13,** redrawn with permission from Meech R.W. et al., *J. Neurophys. 69*: 894–901, © 1993 The American Physiological Society; **3.14,** modified from Anderson P.A., *Comp. Biochem. Physiol. B Biochem. Mol. Biol. 129*: 17–28, © 2001 Elsevier; **3.15,** redrawn from Levitan E.S. et al., *Proc. Natl. Acad. Sci. 84*: 6307–6311, © 1987 NAS, USA; **3.16,** redrawn from Gorman A.L.F. et al., *Fed. Proc. 40*: 2233–2239, © 1981 FASEB; **3.17,** modified from Stent G.S. et al., *Science 200*: 1348–1357, © 1978 AAAS, and Kristan W.B. et al., *J. Comp. Physiol. 94*: 97–119, © Springer Science and Business Media; **3.18,** modified from Nicolls J.G. et al., *Sci. Am. 230(1)*: 38–48, with permission from Nelson H. Prentiss; **3.19 and 3.23,** modified from Kristan W. et al., *Prog. Neurobiol. 76*: 279–327, © 2005 Elsevier; **3.20,** modified from Stent G.S. et al., *Science 200*: 1348–1357, © 1978 AAAS; **3.21 and 3.22,** redrawn, modified from Friesen W.O. et al., *Proc. Natl. Acad. Sci. 73*: 3734–3738, © 1976 NAS, USA.

Chapter 4

Quote, Moore M., reprinted from *Observations*, © 1924 The Dial Press, New York; **4.1,** image provided by Genny Anderson, Santa Barbara City College, used with permission; **4.2,** modified from Kandel E.R., *Cellular basis of behavior*, © 1976 W.H. Freeman & Co, San Francisco; **4.3,** modified with permission from Marinesco S. et al., *J. Neurophysiol. 92*: 2468-2486, © 2004 The American Physiological Society; **4.4,** modified from Mackey S.L. et al., *J. Neurosci. 9*: 4227–4235, © 1989 Society for Neuroscience; **4.5,** modified from Trudeau L.E., *J. Neurophysiol. 70(3):* 1210–1220, © 1993 The American Physiological Society; **4.6,** modified from Stark L.L. et al., *J. Neurosci. 19:* 334–346, © 1999 Society for Neuroscience; **4.7,** modified from Wright W.G., *Bioscience 50:* 883–894, © 2000 American Institute of Biological Sciences; **4.8,** modified from Erixon N.J. et al., *J. Comp. Neurol. 408:* 506–514, © 1999 John Wiley & Sons, Inc.; **4.9 and 4.10,** reproduced, modified from Marinesco S. et al., *J. Physiol. 550:* 241–253, © 2003 Blackwell Publishing, image courtesy of William Wright; **4.11,** modified from Wu J.-Y. et al., *J. Neurosci. 14:* 1366–1384, © 1994 Society for Neuroscience; **4.12,** modified from Tsau Y. et al., *J. Neurosci. 14:* 4167–4184, © 1994 Society for Neuroscience; **4.13,** modified from Zecevic D. et al., *J. Neurosci. 9:* 3681–3689 © 1989 Society for Neuroscience; **4.14,** modified from Briggman K.L. et al., *Science 307:* 896–901, © 2005 AAAS.

Chapter 5

Quote, Frost R., reprinted from *Selected poems,* © 1928 Henry Holt and Company, Inc.; **5.1 and 5.7,** modified from Lear B.C. et al., *Neuron 48:* 221–227, © 2005 Elsevier; **5.2,** modified from Hyun S. et al., *Neuron 48:* 267–278, © 2005 Elsevier; **5.3,** redrawn from Schwartz W.J., *Nature 431:* 751–752, © 2004 Macmillan; **5.4 ,** reproduced with modifications from Renn S.C. et al., *Cell 99:* 791–802, © 1999 Elsevier; **5.5,** redrawn from Hardin P.E., *Curr. Biol. 15:* R714–R722, © 2005 Elsevier; **5.6,** modified from Hall J.C., *Trends Neurosci. 18:* 230–240, © 1995 Elsevier; **5.8,** reproduced from Lin Y. et al., *J. Neurosci. 24:* 7951–7957, © 2004 Society for Neuroscience; **5.9,** modified from Kaneko M. et al., *J. Comp. Neurol. 422:* 66–94, © 2000 John Wiley & Sons Inc.; **5.11,** modified from Young M.W., *Science 288:* 451–453; © 2000 AAAS, and Panda S. et al., *Nature 417:* 329–335, © 2002 Macmillan; **5.12, left,** López-Cortés A., *Arch. Hydrobiol. Suppl. 129:* 245–248, © 1999 E. Schweizerbart, Borntraeger, and Cramer Science Publishers, http://www.schweizerbart.de, image courtesy of Alejandro López Cortés; **5.12, right,** courtesy of J. William Schopf, UCLA (fossil cyanobacterium, Bitter Springs Formation, central Australia, ~850 million years old); **5.13 and 5.15,** modified from Sawyer L.A. et al., *Science 278:* 2117–2120, © 1997 AAAS; **5.14,** modified from Costa R. et al., *Proc. Biol. Sci. 250:* 43–49, © 1992 The Royal Society.

Chapter 6

Quote, Alberti L.B., "La mosche," in *Opere volgari, a cura di Cecil Grayson,* G. Laterza, Bari, Italy, 1960–1973, translated by Mr. James Brusuelas and Dr. Dana F. Sutton, University of California, Irvine (unpubl.); **6.1,** reproduced from Trimarchi J.R. et al., *J. Zool. 235:* 211–222, © 1995 Blackwell Publishing; **6.2 and 6.8,** redrawn, modified from Dickinson M.L., *Curr. Biol. 15:* R1–R5, © 2003 Elsevier; **6.3,** redrawn from Harcombe E.S. et al., *J. Comp. Physiol. A 123:* 271–279, © 1978 Springer Science and Business Media; **6.4,** redrawn from Miller A., *Biology of* Drosophila, © 1950 M. Demerec, Hafner Publishing Co., Inc., after Engel J.E. et al., *J. Comp. Physiol. A 171:* 93–104, © 1992 Springer Science and Business Media; **6.5,** redrawn from Trimarchi J.R. et al., *J. Comp. Physiol. A 176:* 355–364, © 1995 Springer Science and Business Media; **6.6,** reproduced from Phelan P. et al., *J. Neurosci. 16:* 1101–1113, © 1996 Society for Neuroscience, image courtesy of Pauline Phelan; **6.7,** redrawn from Trimarchi J.R. et al., *J. Exp. Biol. 198 (Pt 5):* 1099–1104, © 1995 Company of Biologists; **6.8,** modified from Dickinson M.L., *Philos. Trans. R. Soc. London B 354:* 903–916, © 1999 The Royal Society; **6.9,** modified from Fayyazuddin A. et al., *J. Neurophysiol. 82:* 1916–1926, © 1999 The American Physiological Society; **6.11, 6.12, and 6.14,** modified, reproduced, redrawn from Borst A. et al., *J. Comp. Physiol. A 188:* 419–437, © 2002 Springer Science and Business Media; **6.13,** modified from Zbikowski R., *IEEE Spectrum* November 2005: 46–51, © Bryan Christies, Krapp H.G. et al., *Nature 384:* 463–466, © 1996 Macmillan, and Borst A. et al., *Invertebrate neurobiology* (North and Greenspan, eds.), Chapter 7, © 2007 CSHLP; **6.15,** modified from Egelhaaf M., *Biol. Cybernet. 52:* 195–209, © 1985 Springer Science and Business Media; **6.16,** modified from Frye M.A. et al., *Neuron 32:* 385–388, © 2001 Elsevier; **6.17,** reproduced, modified from van Swinderen B. et al., *Nat. Neurosci. 6:* 579–586, © Macmillan 2003, image courtesy of

Zuse Institute Berlin (ZIB) and Mercury Computer Systems, Inc.; microscopy data courtesy of Department of Genetics, University of Wurzburg, Germany; **6.18,** modified from van Swinderen B. et al., *Nat. Neurosci. 6:* 579–586, © 2003 Macmillan; **6.19,** image courtesy of Jenée Wagner, The Neurosciences Institute; **6.20,** redrawn from Guo A. et al., *Learn. Mem. 4:* 192–204, © 1997 CSHLP.

Chapter 7

Quote, Shakespeare W., *King Lear,* Act IV, Scene VI; **7.1,** modified from Ferveur J.-F., *Behav. Genet. 35:* 279–295, © 2005 Springer Science and Business Media; **7.2,** modified from Amrein H., *Curr. Opin. Neurobiol. 14:* 435–442, © 2004 Elsevier; **7.3,** redrawn from Bray S., *Neuron 39:* 1019–1029, © 2003 Elsevier; **7.4,** reproduced with modifications from Kimura K. et al., *Nature 438:* 229–233, © 2005 Macmillan; **7.5, bottom,** adapted and reprinted from Greenspan R.J. et al., *Annu. Rev. Genet. 34:* 205–232, with permission from Annual Reviews, www.annualreviews.org; **7.6,** redrawn from Kulkarni S.J. et al., *Genetics 118:* 267–285, © 1988 Genetics Society of America; **7.7,** modified from Frye M.A. et al., *Curr. Opin. Neurobiol. 14:* 729–736, © 2004 Elsevier; **7.8,** modified from Todi S.V. et al., *Microsc. Res. Tech. 63:* 388–399, © 2004 John Wiley & Sons, and Demerec M., *Biology of* Drosophila, pp. 382 and 502, © 1950 CSHLP; **7.9,** modified with permission from Kyriacou C.P. et al., *Proc. Natl. Acad. Sci. 77:* 6729–6733, © 1980 NAS, USA; **7.10,** modified from Kyriacou C.P. et al., *Nature 314:* 171–173, © 1985 Macmillan; **7.11,** modified from http://flybrain.neurobio.arizona.edu/Flybrain/html/atlas/schematic/frontschem1.html, © 1995–2000 Flybrain; **7.12,** modified from Ferveur J.F. et al., *Genetica 97:* 73–80, © 1996 Springer Science and Business Media; **7.13,** modified from Tauber E. et al., *Curr. Biol. 13:* 140–145, © 2003 Elsevier; **7.14,** modified from Campesan S. et al., *Genetics 158:* 1535–1543, © 2001 Genetics Society of America.

Chapter 8

Quote, Wittgenstein L., *Tractatus logico-philosophicus,* English translation by C.K. Ogden, © 1922 Routledge and Kegan Paul, London; **8.2,** modified from Amrein H., *Curr. Opin. Neurobiol. 14:* 435–442, © 2004 Elsevier; **8.3,** modified from http://www.cshl.edu/public/SCIENCE/tully.html; **8.4,** modified from http://flybrain.neurobio.arizona.edu/Flybrain/html/atlas/schematic/frontschem1.html, © 1995–2000 Flybrain; **8.5,** image courtesy of Randolf Menzel, used by permission; **8.6,** redrawn from Menzel R. et al., *Learn. Mem. 8:* 198–208, © 2001 CSHLP; **8.7,** modified from Hammer M., *Nature 366:* 59–63, © 1993 Macmillan; **8.9,** redrawn from Faber T. et al., *Nat. Neurosci. 2:* 74–78, © 1999 Macmillan; **8.10,** redrawn with modifications from Giurfa M. et al., *Curr. Opin. Neurobiol. 7:* 505–513, © 1997 Elsevier, and Giurfa M. et al., *Nature 382:* 458–461, © 1996 Macmillan; **8.11,** redrawn with modifications from Giurfa M., *Curr. Opin. Neurobiol. 13:* 726–735, © 2003 Elsevier, and Giurfa M. et al., *Nature 410:* 930–933, © 2001 Macmillan; **8.12,** modified from Giurfa M. et al., *Curr. Opin. Neurobiol. 7:* 505–513, © 1997 Elsevier, and Menzel R. et al., *Anim. Behav. 55:* 139–152, © 1998 Elsevier; **8.13,** redrawn from Mizunami M. et al., *J. Comp. Neurol. 402:* 520–537, © 1998 Wiley-Liss, Inc., a subsidiary of John Wiley & Sons, Inc.; **8.14 and 8.15,** redrawn from Mizunami M. et al., *J. Comp. Neurol. 402:* 501–519, © 1998 Wiley-Liss, Inc., a subsidiary of John Wiley & Sons, Inc.

Chapter 9

Quote, Twain M., *Following the Equator, A journey around the world,* Chapter XXVII, title page, © 1897, American Pub. Co., Hartford, Connecticut.

Index

Entries followed by f denote figures.